D1632091

som 969

Blackwell
Science

Building LifePlans

Building Services
Component Life Manual

© 2001 Building Performance Group Limited

Blackwell Science Ltd
Editorial Offices:
Osney Mead, Oxford OX2 0EL
25 John Street, London WC1N 2BS
23 Ainslie Place, Edinburgh EH3 6AJ
350 Main Street, Malden,
    MA 02148 5018, USA
54 University Street, Carlton,
    Victoria 3053, Australia
10, rue Casimir Delavigne, 75006 Paris, France

Other Editorial Offices:

Blackwell Wissenschafts-Verlag GmbH
Kurfürstendamm 57
10707 Berlin, Germany

Blackwell Science KK
MG Kodenmacho Building
7–10 Kodenmacho Nihombashi
Chuo-ku, Tokyo 104, Japan

Iowa State University Press
A Blackwell Science Company
2121 S. State Avenue,
Ames, Iowa 50014–8300, USA

First published 2001

Designed and set in ITC News Gothic
by Paul Jones Associates
Printed and bound in Great Britain by
MPG Books, Bodmin, Cornwall.

The Blackwell Science logo is a trade mark of
Blackwell Science Ltd, registered at the United
Kingdom Trade Marks Registry

DISTRIBUTORS

Marston Book Services Ltd
PO Box 269
Abingdon
Oxon OX14 4YN
(Orders: Tel: 01235 465500
                  Fax: 01235 465555)

USA
Blackwell Science, Inc.
Commerce Place
350 Main Street
Malden, MA 02148 5018
(Orders: Tel: 800 759 6102
                        781 388 8250
                  Fax: 781 388 8255)

Canada
Login Brothers Book Company
324 Saulteaux Crescent
Winnipeg, Manitoba R3J 3T2
(Orders: Tel: 204 224-4068)

Australia
Blackwell Science Pty Ltd
54 University Street
Carlton, Victoria 3053
(Orders: Tel: 03 9347 0300
                  Fax: 03 9347 5001)

A catalogue record for this title
is available from the British Library

ISBN 0-632-05887-0

For further information on
Blackwell Science, visit our website:
www.blackwell-science.com

The research for this manual was commissioned by Defence Estates, an executive agency of the UK
Ministry of Defence. It was compiled by durability specialists, Building Performance Group Limited, with
the assistance of BSRIA, AGS Consulting and David Miles and Partners. The principal authors were Gary
Moss, Julian King and Andrew Saville.

This publication has been sponsored by Building LifePlans Ltd which has adopted the data to support its
long term insurance for commercial buildings.

All technical enquiries should be directed to:

Building Performance Group Ltd
Grosvenor House
141–143 Drury Lane
London WC2B 5TS

Tel: 020 7240 8070
Fax: 020 7836 4306

E-mail: enquiries@bpg-uk.com
Web Site: www.bpg-uk.com

The authors acknowledge the license of Defence Estates to reproduce copies of this Manual for UK
government use, under contract DEO11/4314.

Contents of the Manual

Foreword

The benefits of whole life costing of building services

Over the next 20 years the UK will build four million new homes and millions of square feet of office space, industrial and retail units. Planners, lobbyists and ministers will argue and spill ink over their environmental impact, and some might even ask whether we need them all. But how many people will think about how long they will last? Who wants to know the costs of maintenance, repair, and even replacement of the services within?

Increasingly, informed clients are asking these questions. They want to know the costs of owning buildings over their whole life. They do not want the cheapest build cost at any price in maintenance bills later. Clients want predictable costs of ownership and will pay extra at the outset to reduce operating costs, if they are given realistic and carefully prepared whole life cost predictions. The recent Ministry of Defence *Building Down Barriers* project achieved a 16% reduction in predicted whole life costs, with small increases in capital cost.

For some time now there has been a lack of real data with which to predict whole life costs. Sometimes, whole life costs are based on a forward projection of historic performance for which data is plentiful. But that is changing as new tools become available to help those who really want to develop whole life cost models.

Predicting whole life cost requires knowledge about the long-term performance, or durability, of a material, component or element. It requires an understanding of how and when the component is likely to fail under a range of exposure conditions. The first systematic source of such data was the HAPM Component Life Manual, developed in 1992 for the social housing sector. It predicts the performance of about 500 domestic building components. The BPG Building Fabric Component Life Manual was published in 1999 and covers some 130 components used in commercial and industrial buildings. In addition to these sources of data, a new British and International Standard addresses the issue of whole life costing, and the Construction Clients Forum have recently published *A Client's Guide to Whole Life Costing*. These two provide a methodology and advice to clients who want whole life cost predictions for their projects, respectively.

A significant gap in all this has been the performance of building services. Often accounting for 50–60% of the capital cost of a building and a significantly higher proportion of the running costs, they also require regular maintenance, and often need to be replaced at least once during the life of the structure. (Guidance on maintenance has recently been published by the CIBSE, as the Guide to Ownership, Operation and Maintenance). Reliable building services which will last as long as they are meant to can provide significant real savings to building operators, as well as operating more efficiently and saving energy (and climate change levy) in the process. But this can only be achieved if robust guidance is made available on the whole life performance of building services

This new publication addresses that need. It gives hard data on the likely performance of a range of 20 major services components, enabling consultants or client advisors to predict their whole life costs. It gives guidance on the issues which influence the durability of services components, and provides performance benchmarks for different levels of specification. It will extend the scope of whole life costing into the field of building services. It will help to reduce the costs and the impacts of all that new development. And, because it applies to building services, it can help to reduce the whole life costs when it is time to replace old plant in existing buildings too.

Dr Hywel Davies
Chairman, BSI sub-committee for Durability
CIBSE Research Manager

Acknowledgements

The authors are indebted to the many individuals and organisations that assisted with the preparation of this manual, and without whom the task would have been considerably more difficult to complete. Particular thanks are due to the numerous manufacturers, suppliers, trade and research organisations who willingly contributed their time, data and support to this project.

Much of the original source material for this manual was assembled under a DTI funded Teaching Company research project undertaken by Building Performance Group and the University of the West of England. The valuable contribution made by this project and the principal researcher, Ed Bartlett, is greatly appreciated.

Introduction

This Manual sets out the interim findings of an ongoing programme of research into the longevity of building services plant and was prepared under the sponsorship of Defence Estates, an executive agency of the UK Ministry of Defence. It forms the third in a series of component life manuals compiled by durability specialists, Building Performance Group and arose from an urgent need for robust comparative data on the longevity of mechanical and electrical plant and equipment used in typical commercial, industrial and public building types.

The first durability manual, the *HAPM Component Life Manual,* schedules "insured lives" and maintenance requirements for in excess of 500 domestic building components, materials and assemblies. First published in 1992, the HAPM Manual was prepared to underpin the Housing Association Property Mutual (HAPM) 35 year latent defects insurance scheme for housing associations. HAPM insures components in respect of premature failure and it is therefore necessary to assign "insured lives" to those components with a life expectancy of less than the 35 year insurance period. Building Performance Group subsidiary, Construction Audit Limited, was therefore commissioned by HAPM to prepare detailed schedules of "insured lives" and maintenance requirements for common domestic building components.

In 1998 Defence Estates commissioned a companion volume to the HAPM manual to provide further data on a range of components typically found in commercial and industrial building types. The BPG Building Fabric Component Life Manual sets out insured lives and maintenance requirements for some 130 components including floor screeds and toppings, cladding systems and internal partitions.

This third publication follows the same basic format and structure as the previous manuals. This includes the concept of "insured" lives for components to support the Building LifePlans' (BLP) insurance product. These represent a somewhat cautious prediction of the actual service life. Consequently, readers may sometimes wish to apply their own adjustment factors before using the data for other purposes such as maintenance scheduling and whole life costing exercises.

The list of building services systems and components covered by this Manual has been selected to include those that are commonly used in commercial, industrial and public building types and which are not covered by the existing HAPM Component Life Manual. Whilst project funding limited the scope to 20 key components for this publication, further research, funded by the DETR Partners in Innovation programme, is under way in preparation for an enlarged and updated second edition in the summer of 2002.

It is important to note that this manual was not conceived as a comprehensive, in-depth guide to the specification, detailing, installation, commissioning or maintenance of building services plant. Rather, it aims to make explicit, in a straightforward, concise manner, the different levels of plant specification that are available, their expected longevity, the typical inspection and maintenance regimes that are necessary to support that longevity, and the often complex range of factors that combine to determine their durability in practice. The expected readership ranges from the surveyor or facilities manager with only a cursory knowledge of building services installations, to the specialist services consultant or design engineer, with a high level of expert knowledge. Throughout the text, extensive references have been made to the plethora of guidance available within the industry, from organisations such as the British Standards Institute (BSI), the Building Services Research and Information Association (BSRIA) and the Chartered Institution of Building Services Engineers (CIBSE). Readers should refer to this source material for more detailed guidance and information.

Use of the Manual

Structure

The Manual is divided into a number of Sections, each representing a major building services system. The structure is broadly based on the Uniclass and CAWS works sections.

Section 1:   Piped supply systems
Section 2:   Mechanical heating/cooling/refrigeration systems
Section 3:   Ventilation and air conditioning systems
Section 4:   Electrical supply/power/lighting systems
Section 5:   Communications/security/control systems
Section 6:   Transport systems

Systems, Components and Component Sub-types

The durability studies contained in this manual focus on the major components, or plant items that make up the above building services systems. A total of 20 common building services components have been included in this first edition. Further components will be included in future editions.

Each of the 20 components included in the manual is further divided into a number of subtypes which represent the most common types, materials and/or configurations available in the UK. For each component subtype, a series of distinct specification benchmarks, or quality levels, is provided and each description is assigned a life assessment (ranging from 5 to 35+ years). A further life class (designated "U" for 'unclassified') is assigned to components that fail to comply with relevant British/European Standards or are unsuitable for the purpose specified, or where there is insufficient information provided to enable a life to be allocated.

It is important to note that this Manual in no way seeks to limit the choices of building services components available to specifiers. It may be quite valid to select a short life component where access for maintenance or replacement will be straightforward and higher costs in use are acceptable. The need for replacement may be dictated by factors unconnected with the failure of a component, such as non-availability of spare parts or a desire to improve standards. Nevertheless, the ability to differentiate between qualities of component in order to assign a longer life to those which can be identified as having enhanced durability characteristics is considered to be a powerful tool for specifiers. Component descriptions have therefore been ranked in order of durability, which could also be termed "quality" in a particular and a limited sense.

An introduction page is provided for each component in the manual, outlining the extent of the coverage and providing a bibliography of all British and European standards and other references cited.

The Life Assessments

To enable this Manual to be used alongside the existing HAPM and BPG Component Life Manuals the life assessments quoted are analogous to "insurance lives", which represent somewhat cautious assessments of durability. The lives are assigned in 5-year bands up to a maximum of 35+, which represents components expected to last in excess of 35 years. The insurance lives assume compliance with good practice in design and installation, a normal amount of maintenance, and typical exposure and usage conditions.

Location

Where the service life of a component is likely to vary depending on its location in the building, (e.g. internal, external), alternative life assessments are given for different locations.

Adjustment Factors

The life assessments given in the manual are based on 'normal' environmental conditions, use patterns and protective measures. Where the service life is likely to be affected by changes in these factors, positive and negative adjustment factors are provided to enable lives to be amended.

Definitions for environmental conditions which are used throughout the Manual have been included in an appendix for guidance.

Assumptions

In assigning life assessments to components certain general assumptions have been made throughout the manual in addition to specific assumptions which relate to particular components.

General assumptions that apply throughout include installation in accordance with manufacturers' directions, good practice, relevant Codes of Practice and British/European Standards and the use of appropriate design details. In addition, where a third party assurance certification scheme (e.g. British Board of Agrément (BBA) certificate) applies, installation is assumed to comply with the conditions of the certificate.

Specific assumptions, highlighting particular facets of system design or available guidance on installation and commissioning, have been made in estimating the service lives of specific components. These may reflect, for example, compliance with a trade standard for installation, or ensuring that incompatible materials are adequately separated. Such assumptions are stated on the relevant pages in the Manual.

Non-compliance with these assumptions may have a negative effect on component durability.

Inspection

Guidance on the minimum inspection activities and intervals (based on good practice guidance or statutory provisions) is provided for each component subtype in the manual. Life assessments are assigned on the basis that the defined inspection routines, or other suitable routines as approved by the manufacturer, are adhered to and that any necessary maintenance or corrective work identified during the inspection is carried out.

Maintenance

Life assessments are assigned to components on the assumption that a certain minimum level of maintenance will be carried out. Each data sheet specifies the minimum maintenance activities and intervals, based on good practice guidance and typical manufacturers' requirements. Any specific manufacturer's maintenance requirements must also be adhered to. It should be noted that the maintenance schedules are not intended to be exhaustive, but to indicate the typical level of work required to maintain the performance of the component.

Notes

Where appropriate, additional notes are provided on matters such as further sources of information, incompatibilities, unsuitable applications or materials.

Key Failure Modes and Key Durability Issues

Lists of key failure modes and key durability issues are provided for each component, to explain the reasoning behind the life assessments and to highlight the key determinants of durability in practice. Key failure modes represent the mechanisms by which the component is most likely to fail. Key durability issues direct readers to the issues that are most likely to influence durability.

Origins of the durability data

The data in this manual have been assembled by Building Performance Group using a rigorous research process developed from the company's 10 years' experience in the field of building component durability. The durability assessment process was originally developed to provide insured life assessments for the HAPM latent defects insurance scheme on its inception in 1990. It has since been refined through a four year collaborative research programme with the University of the West of England and through research and consultancy work for clients such as the National Housing Federation, Building Research Establishment, Defence Estates and British Standards Institute.

For each component included in the manual, service life and failure data has been assembled from a diverse range of sources including, where relevant:

- Manufacturers and suppliers

- Manufacturer and trade associations

- Professional and research organisations such as CIBSE, BSRIA, ASHRAE

- Test houses and certification bodies

- Building services consultants

- Building owners and maintenance managers

- British and European standards

- Published research and conference papers

- Technical press

- Existing published sources of component life data

Wherever possible the above data has been supplemented by feedback from the performance of buildings in use and on the most commonly encountered causes of deterioration and failure.

Analysis of this data has enabled a series of 'quality benchmarks', or levels of performance, to be identified for each building component under consideration. Life assessments and statements of minimum required maintenance have been assigned to each benchmark.

The prediction of building component lives is not an exact science – it is acknowledged that the data presented in this manual can be no more than an informed estimate based on the information available at the time of writing. The authors keep under regular review changes in British and European Standards, and developments by testing and research organisations and product manufacturers. The component descriptions and life assessments will be reviewed periodically to take account of such developments and any necessary changes will be incorporated into future editions. The authors welcome any comments or feedback from users of the manual.

About the authors

Building Performance Group is a multi-disciplinary consultancy with an international reputation for its work in the field of building component durability. Its publications on the subject include the groundbreaking *HAPM Component Life Manual*, the *BPG Building Fabric Component Life Manual*, the *Defence Estates Guide to Durability Auditing* and a BRE published *Defects Avoidance Manual*. The company is represented on a number of industry research and standards committees and has drafted parts 1 and 3 of the new International Standard on service life planning for buildings.

As technical auditor to the Housing Association Property Mutual and Building LifePlans latent defects insurance schemes, BPG has unique access to robust data on the performance in use of many thousands of buildings. This data is supplemented by the company's extensive experience in the fields of condition surveys and inspections, defects investigations, maintenance planning and refurbishment of a broad range of building types.

Factors affecting the longevity of building services plant

Obsolescence and durability

Whilst this manual sets out typical replacement intervals based on durability or functional performance, it is important to recognise that in practice plant replacement may also be driven by obsolescence, or by an inability to satisfy changing requirements. BS ISO 15686:Part 1 defines three types of obsolescence: functional, technological and economic.

Typical examples of obsolescence in building services plant include:

Functional:	changing functional requirements or patterns of building use, e.g. more/less cooling required, or required in different locations, new plant available with enhanced functionality
Technological:	advances in technology resulting in new plant with enhanced performance or functionality, e.g. new computerised control systems with greater efficiency and functionality
Economic:	development of plant with greater efficiency, lower energy and maintenance costs, e.g. condensing boilers

Since it is rarely possible to predict future obsolescence with any degree of accuracy, its effects have been excluded from the durability schedules in this manual. However, in some cases it may be pertinent for users to build in provisions for anticipated future obsolescence, e.g. provision of access for the future replacement of large plant items.

Factors affecting plant longevity

In practice, the durability or service life of a particular plant item will be the result of a complex interaction between several influencing agents, or factors, that will vary according to its inherent quality, location, operation and maintenance regime. A summary of these factors, adapted from Table E.1 of BS ISO 15686-1, is provided below.

Agents	Factor	Examples
Agents related to inherent quality characteristics	A Quality of components	Manufacture, storage, transport, materials, protective coatings
	B Design/detailing	Incorporation into the building, detailing, system design, interfaces
Agents related to environment	C Installation/workmanship	Site management, standard of workmanship, climatic conditions during installation
	D Indoor environment	Aggressiveness of environment, ventilation, condensation
	E Outdoor environment	Location of building, micro and macro environment, sheltering, pollution levels, weathering factors
Agents related to operation conditions	F In-use conditions	Commissioning, hours/frequency of use, mechanical impact, category of users, wear and tear
	G Maintenance	Quality and frequency of inspection and maintenance, accessibility for maintenance

It should be noted that the factors affecting the longevity of building services plant often differ significantly from those affecting building fabric components. Plant items are highly dependent upon the effective design of the system in which they operate. Many of their components are dynamic in nature and are consequently dependent on timely and appropriate maintenance. Their longevity is also affected greatly by the way in which they are used. In many cases, manufacturers quote longevity in terms of 'hours of operation' rather than 'years of service'. These and other factors affecting the longevity of building services plant are discussed in more detail below.

Quality of components

This manual is founded on the principle that for any given plant item it is possible to identify a number of defined 'benchmarks' of specification or quality in terms of likely service life performance. These benchmarks relate to the materials used to construct the item, the way in which it is designed and manufactured, and the standards and codes of practice with which it complies. For each plant item included in this manual, these factors are identified and a number of generic specification descriptions are provided, each of which is assigned a life assessment. The number of generic descriptions varies from component to component and reflects the different levels of specification available within the UK. Service life assessments are assigned on the assumption that the higher the level of specification, the longer the product is likely to last (assuming that all other operating conditions remain the same).

Typical quality issues cited in the component durability schedules include:

- Compliance with British/European standards and with levels of performance defined within standards (e.g. minimum usage cycles)
- Availability of third party certification to verify performance (e.g. BSI Kitemarking, Agrément Certification, recognised product certification schemes).
- Evidence of manufacturing quality control/checks, e.g. compliance with ISO 9000 and other recognised quality schemes
- Materials from which the component is constructed, and method of construction (where relevant to durability)
- Method(s) of protection of the component, e.g. surface protection against corrosion
- Configuration/assembly of the components, e.g. geared/gearless drives, direct/belt driven fans and pumps
- Potential for future overhauling or re-manufacture, e.g. re-greasable bearings, rewindable motors.

Further issues affecting component quality, such as transportation, storage and protection prior to installation, are discussed in the notes following each component durability schedule.

Design/detailing

Appropriate design and detailing, both of the component under consideration and of the system in which it operates, are fundamental to the longevity of building services plant. Key design/detailing issues cited in the durability schedules include:

- System design, e.g. interfaces and interaction between components, compatibility of system components, provision of necessary safety and control devices
- Protection, e.g. against adverse environments, temperature extremes, overloading
- Fitness for purpose, e.g. selection of plant that is adequately sized, suitable for the intended use, location, operating conditions, loading, hours of operation etc.
- Incompatibilities, e.g. isolation of dissimilar metals in order to prevent galvanic corrosion
- Key design and detailing issues are highlighted in the durability schedules, along with references to relevant standards and sources of good practice guidance.

Installation/workmanship

The service life assessments in the manual are based on the assumption that installation is carried out to appropriate standards of workmanship and in accordance with manufacturer's instructions, relevant standards and other good practice guidance. Key issues are highlighted in the durability schedules, along with guidance on essential commissioning activities such as flushing and cleaning, load testing and adjustment of systems prior to handover of the completed facility.

Indoor environment

Key issues relating to the indoor environment include protection against temperature extremes, humidity and corrosive environments, air quality and impact or abrasion.

Outdoor environment

The outdoor environment to which a component is exposed can have a significant effect on its long term performance. Corrosion of metal components due to moisture and acidic pollutants is a key concern with building services plant, although wind driven particulates and exposure to ultra-violet radiation can also be significant. Negative adjustment factors are provided in the durability schedules to factor component lives in particularly harsh environments such as polluted or marine environments. Similarly, positive adjustment factors are provided to reflect enhanced surface protection or high levels of specification.

In-use conditions

The way in which plant items are used, e.g. frequency and duration of operation, number of cold-starts, can have a significant effect on longevity. Unless stated otherwise, an average 12 hour day, 5 day week is assumed for plant items having moving parts. Where longer working hours are likely to significantly affect durability, adjustment factors are provided. It is also assumed that plant will be operated within the intended design parameters and that no overloading or excessive operating demands will be imposed.

Maintenance

The type of maintenance activities and their frequency are critical to the longevity of building services plant and equipment due to the preponderance of moving parts that require testing, cleaning, lubrication, protection, adjustment, repair and replacement at varying intervals in order for the plant item to perform effectively. Failure to adhere to specified maintenance regimes can lead to accelerated deterioration, e.g. due to wear, corrosion, overloading, or overheating. It can also lead to a decline in performance and efficiency of the plant and, in extreme cases, total plant failure. In many cases, the timing of maintenance and replacement activities will be driven by periodic (statutory or non-statutory) inspections or by condition monitoring. Adherence to specified inspection and monitoring routines can therefore be as important as the maintenance activities themselves.

Provision of adequate access and space to enable inspection and maintenance activities to be carried out effectively is also a key consideration; for further guidance see Defence Works Functional Standard: Design and Maintenance Guide 08.

A note about reliability

A great deal of published material exists on the subject of plant reliability and failure prediction. Reliability theory is concerned more with the probability of failure or the 'mean time to failure' than with the notion of durability or service life explored in this manual. Whilst it is acknowledged that a significant body of plant reliability data exists, the level of skill required to correctly interpret and use the data is often restricted to specialist engineers. The decision has therefore been taken to limit the data in this manual to general service life assessments rather than attempting to assemble detailed reliability data.

Further reading

The following publications, which have been used in the preparation of this manual, contain more detailed guidance on the design, operation and maintenance of building services plant and equipment. Further detailed bibliographies are provided as part of each component study in this manual.

Guide to ownership, operation and maintenance of building services CIBSE Guide
(London: Chartered Institution of Building Services Engineers) (2000)

Jonsson A and Lindgren S,
The longevity of building services installations.
Stage 1: Inventory of operating experiences for estimating longevity
Report 811662–4/80s, Swedish Building Research Council BSRIA
Translation No 252 (Bracknell: Building Services Research & Information Association) (1983).

Kirk S and Dell'Isola A,
Life cycle costing for design professionals
(New York: McGraw Hill) (1995).

Buildings and constructed assets – Service life planning: Part 1: General principles
BS Handbook HB10141 (London: British Standards Institute) (1998)

HAPM Component Life Manual
Construction Audit Ltd. (London: E & FN Spon) (1992)

Maintenance of mechanical services
(London: Department of Education & Science/The Stationery Office) (1990)

Maintenance of electrical services
(London: Department of Education & Science/The Stationery Office) (1992)

HVAC Systems and equipment ASHRAE Handbook
Chapter 33: Owning and operating costs
(Atlanta GA: Americal Society of Heating, Refrigeration & Air Conditioning) (1995)

Standard maintenance specification for mechanical services in buildings: volumes 1 to 5
(London: Heating & Ventilating Contractors' Association) (various dates)

Building services maintenance management CIBSE Technical Memorandum TM17
(London: Chartered Institution of Building Services Engineers) (1994)

CIBSE Commissioning codes
(London: Chartered Institution of Building Services Engineers) (various dates)

BSRIA Application guides
(Bracknell: Building Services Research and Innovation Association) (various dates)

List of Components

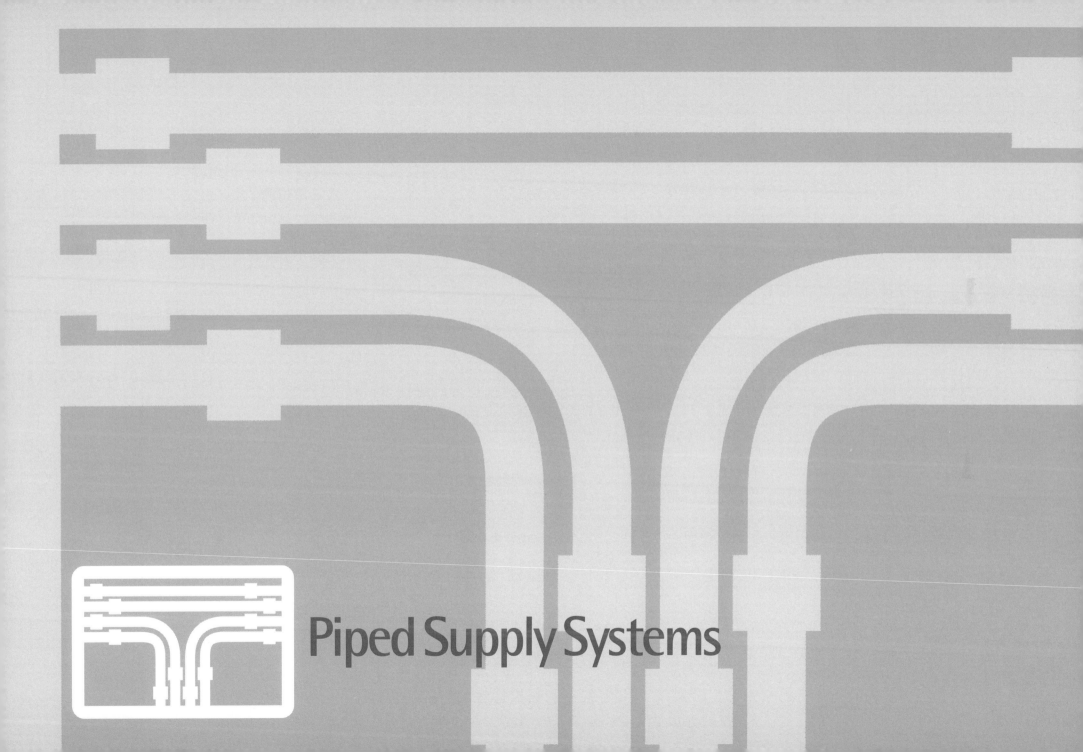

Piped Supply Systems

DISTRIBUTION PIPEWORK

Scope

This section provides data on distribution pipework commonly used in commercial and light industrial applications. Pipework for heavy industrial/manufacturing processes is beyond the scope of this study. Domestic pipework is covered in the HAPM Component Life Manual (see introduction to this manual). Steam pipe systems are also beyond the scope of this manual.

The following component sub-types are included within this section:

Standards cited

BS 143 & 1256:1986	Specification for malleable cast iron and cast copper alloy threaded pipe fittings.
BS 1224:1970	Specification for electroplated coatings of nickel and chromium.
BS 1387:1985	Specification for screwed and socketed steel tubes and tubular and for plain end steel tubes suitable for welding or for screwing to BS 21 pipe threads
BS 1965: Part 1:1963	Specification for butt-welding pipe fittings for pressure purposes. Carbon steel.
BS 3412:1992	Methods of specifying general purpose polyethylene materials for moulding and extrusion.
BS 3505:1986	Specification for unplasticised polyvinyl chloride (PVC-U) pressure pipes for cold potable water.
BS 3601:1987	Specification for carbon steel pipes and tubes with specified room temperature properties for pressure purposes.
BS 4127:1994	Specification for light gauge stainless steel tubes, primarily for water applications.
BS 4346	Joints and fittings with unplasticized PVC pressure pipes.
BS 4504: Part 3:1989	Circular flanges for pipes, valves and fittings (PN designated). Steel, cast iron and copper alloy flanges
BS 4508:	Thermally insulated underground pipelines (various parts).
BS 4825: Part 1:1991	Stainless steel tubes and fittings for the food industry and other hygienic applications. Specification for tubes.
BS 5114:1975	Specification for performance requirements for joints and compression fittings for use with polyethylene pipes.
BS 5255:1989	Specification for thermoplastics waste pipe and fittings
BS 5391: Part 1:1976	Specification for ABS pressure pipe Pipe for industrial uses.
BS 5392 Part 1:1976	Specification for acrylonitrile – butadiene – styrene (ABS) fittings for use with ABS pressure pipe. Fittings for use with pipe for industrial uses
BS 5480:1990	Specification for glass reinforced plastics (GRP) pipes, joints and fittings for use for water supply or sewerage.
BS 6920 Part 1:1996	Suitability of non-metallic products for the use in contact with water intended for human consumption with regard to their effect on the quality of the water. Specification.

Standards cited (continued)

BS 7291	Thermoplastics pipes and associated fittings for hot and cold water for domestic purposes and heating installations in buildings (various parts).
BS 7361 Part 1:1991	Cathodic protection. Code of practice for land and marine applications.
BS 7572:1992	Code of practice for thermally insulated underground piping systems.
BS 7838: 1996	Corrugated stainless steel semi-rigid pipe and associated fittings for low pressure gas pipework to 28mm.
BS 8313:1997	Code of practice for accommodation of building services in ducts
BS EN 253:1995	Preinsulated bonded pipe systems for underground hot water networks. Pipe assembly of steel service pipes, polyurethane thermal insulation and outer casing of high density polyethylene.
BS EN 448:1995	Preinsulated bonded pipe systems for underground hot water networks. Fittings assemblies of steel service pipes, polyurethane thermal insulation and outer casing of high density polyethylene.
BS EN 488:1995	Preinsulated bonded pipe systems for underground hot water networks. Steel valve assemblies for steel service pipes, polyurethane thermal insulation and outer casing of polyethylene.
BS EN 489:1995	Preinsulated bonded pipe systems for underground hot water networks. Joint assembly for steel service pipes, polyurethane thermal insulation and outer casing of polyethylene.
BS EN 545:1995	Ductile iron pipes, fittings, accessories and their joints for water pipes – Requirements and test methods.
BS EN 969:1996	Ductile iron pipes, fittings, accessories and their joints for gas pipelines – requirements and test methods.
BS EN 1057:1996	Copper and copper alloys – Seamless, round copper tubes for water and gas in sanitary and heating applications.
BS EN 1254	Copper and copper alloys – plumbing fittings (various parts).

Other references/information sources

BS 1560: Part 3:1989	Circular flanges for pipes, valves and fittings (class designated). Steel, cast iron and copper alloy flanges.
BS 3506:1969	Unplasticized PVC pipe for industrial purposes.
BS EN 751	Sealing materials for metallic threaded joints in contact with 1st, 2nd and 3rd family gases and hot water (various parts).
BS EN 1775:1998	Gas supply – gas pipework in buildings – maximum operating pressure =<5 bar – functional requirements.
BRE Digest 83	Plumbing with stainless steel
BSRIA	Application Guide: AG2/89 The commissioning of water systems in buildings (1989)
BSRIA	Application Guide: AG8/91 Pre-commission cleaning of water systems (1991)
CIBSE Guide B:1986	Installation and equipment data – B1 Heating, B2/B3 Ventilation and air conditioning (1986)
CIBSE	Commissioning code W:1994 – Water distribution systems.
HVCA	Standard Maintenance Specification for Mechanical Services in Buildings – Vols 1 to 5 (1990–92)
Institute of Plumbing:	Plumbing engineering services design guide (withdrawn; revision due late 2000)
PSA/DEO MEEG 1/03	1987 Water treatment
PSA/DEO MEEG 1/05	1989 Hot and cold water services
PSA/DEO TICE 140	1979 Treatment of potable water
PSA/DEO MEEG 4/03	1989 Water treatment

Water Supply (Water Quality) Regulations: 1989 (as amended by the Water Supply (Water Quality) (Amendment) Regulations 1991

DISTRIBUTION PIPEWORK

COLD WATER PIPEWORK

YEARS	DESCRIPTION	INSPECTION	MAINTENANCE
35+	Stainless steel pipework for water applications to BS 4127. Fittings to BS 4825.	Long cycle inspection for external corrosion/ damage and joint/fixing integrity.	Generally, pipe systems require little or no maintenance if correctly specified. Any maintenance will be in the form of periodic repainting where materials such as cast iron and some types of plastic (to maintain UV protection) are exposed to the elements or are installed in a corrosive setting.
35+	Galvanised mild steel pipework to BS 1387, medium and heavy grade. Screwed malleable iron fittings to BS 143 & 1256. Butt welding fittings to BS 1965: Part 1 with flanges to BS 4504.		
35+	Chrome plated copper pipework to BS EN 1057 – R250 (thin wall, previously Table X) with seamless chromium plated finish to BS 1224, service condition 2. Fittings to BS EN 1254: Parts 1 & 2.		
35+	Copper pipework to BS EN 1057 – R250. (thin wall, previously Table X). Fittings to BS EN 1254: Parts 1 & 2.		
35+	Ductile iron pipework to BS EN 545: 1994 Fittings as below		
35	UPVC pipework to BS 3505. Fittings to BS 4346.		
35	Thermoplastic pipework including polybutylene, cross linked polyethylene, chlorinated polyvinyl chloride, and co-polymer polypropylene pipe to BBA approval. and to BS 7291. Fittings as above and to BS 5114 for polyethylene pipe.		
30	ABS plastic pipework to BS 5391. Fittings to BS 5392.		
30	BBA or other 3rd party certified plastic/ metal composite pipework and fittings.		
15	GRP pipework to BS 5480, BS 6920:Part 1 and to WRc approval. Fittings as above		
U	Unclassified, i.e. pipework/fittings, not to above standards.		

HOT WATER PIPEWORK

YEARS	DESCRIPTION	INSPECTION	MAINTENANCE
35+	Mild steel pipework to BS 1387 – medium grade, galvanised. Fittings: (as cold water).	Long cycle inspection for external corrosion/ damage and joint/fixing integrity.	Generally, pipe systems require little or no maintenance if correctly specified. Any maintenance will be in the form of periodic repainting where materials such as cast iron and some types of plastic (to maintain UV protection) are exposed to the elements or are installed in a corrosive setting.
35+	Copper pipework to BS EN 1057 – R250 (thin wall, previously Table X). Fittings to BS EN 1254: Parts 1 & 2.		
35+	Copper pipework to BS EN 1057 – R250 (thick wall, previously Table Y) (difficult access areas). Fittings to BS EN 1254: Parts 1 & 2.		
35+	Copper pipework to BS EN 1057 (previously Table Z) (straight runs only, not bent or silver brazed). Fittings to BS EN 1254: Parts 1 & 2.		
35+	Plastic coated copper pipework to BS EN 1057 with seamless polythene coating to BS 3412. Fittings to BS EN 1254: Parts 1 & 2.		
35+	Chrome plated copper pipework to BS EN 1057 (previously Tables X & Y) with seamless chromium plated finish to BS 1224, service condition 2. Fittings to BS EN 1254:Parts 1 & 2.		
35	Thermoplastic pipework including polybutylene, cross linked polyethylene, chlorinated polyvinyl chloride, and co-polymer polypropylene pipe to BBA approval.and to BS 7291. Fittings: as above.		
U	Unclassified, i.e. pipework/fittings, not to above standards.		

	HEATING PIPEWORK		
YEARS	DESCRIPTION	INSPECTION	MAINTENANCE
35+	Galvanized mild steel pipework to BS 1387 – medium & heavy grade. Fittings: as cold water.	Long cycle inspection for external corrosion/ damage and joint/fixing integrity.	Generally, pipe systems require little or no maintenance if correctly specified. Any maintenance will be in the form of periodic repainting where materials such as cast iron and some types of plastic (to maintain UV protection) are exposed to the elements or are installed in a corrosive setting.
35+	Black mild steel pipework to BS 1387 – medium & heavy grade. Fittings: as cold water.		
35+	Black mild steel pipework to BS 3601 grade seamless for large diameter (option where above is not available). Fittings: as cold water.		
35+	Copper pipework to BS EN 1057 (thin wall, previously Table X). Fittings to BS EN 1254: Parts 1 & 2.		
35+	Thermoplastic pipework including polybutylene, cross linked polyethylene, chlorinated polyvinyl chloride, and co-polymer polypropylene pipe to BBA approval and to BS 7291. Fittings: as above.		
30	BBA or other 3rd party certified plastic/metal composite pipework and fittings.		
U	Unclassified, i.e. pipework/fittings, not to above standards.		

Adjustment factors

Installed in adverse (but not severely corrosive) environments: –5 years
Not sleeved through walls: –5 years
Note: The above factors are not cumulative: the factor that is the largest should be applied.

DISTRIBUTION PIPEWORK *(continued)*

FOOD AND HYGIENIC APPLICATIONS PIPEWORK

YEARS	DESCRIPTION	INSPECTION	MAINTENANCE
35+	Stainless steel pipework to BS 4825, part 1, BS 4127. Fittings: as previous page.	Long cycle inspection for external corrosion/ damage and joint/fixing integrity.	Generally, pipe systems require little or no maintenance if correctly specified. Any maintenance will be in the form of periodic repainting where materials such as cast iron and some types of plastic (to maintain UV protection) are exposed to the elements or are installed in a corrosive setting.
35+	Copper pipework to BS EN 1057 (previously Table X &Y). Fittings to BS EN 1254:Parts 1 & 2.		
35+	Chrome plated copper pipework to BS EN 1057 (previously Tables X & Y) with seamless chromium plated finish to BS 1224, service condition 2. Fittings to BS EN 1254:Parts 1 & 2.		
35	Thermoplastic pipework including polybutylene, cross linked polyethylene, chlorinated polyvinyl chloride and co-polymer polypropylene pipe to BBA approval and to BS 7291. Fittings: as above.		Clean down externally and flush/sterilise internally as use/process requires.
35	UPVC pipework to BS5255. Fittings: as above.		
30	ABS plastic pipework to BS 5391. Fittings to BS 5392.		
15	GRP pipework to BS 5480, BS 6920:Part 1 and to WRC bylaws. Fittings: as above.		
U	Unclassified, i.e. pipework/fittings, not to above standards.		

CONDENSER WATER PIPEWORK

YEARS	DESCRIPTION	INSPECTION	MAINTENANCE
35+	Mild steel pipework to BS 1387 – heavyweight – galvanised. Fittings: as cold water.	Long cycle inspections for external corrosion and joint/fixing integrity.	Generally, pipe systems require little or no maintenance if correctly specified. Any maintenance will be in the form of periodic repainting where materials such as cast iron and some types of plastic (to maintain UV protection) are exposed to the elements or are installed in a corrosive setting.
35+	Carbon steel pipework to BS 3601 – galvanised. Fittings to BS 1965, with flanges to BS 4504.		
U	Unclassified, i.e. pipework/fittings, not to above standards.		

CHILLED WATER PIPEWORK

YEARS	DESCRIPTION	INSPECTION	MAINTENANCE
35	Black mild steel pipework to BS 1387 – heavyweight. Screwed malleable iron fittings to BS 143 & BS 1256. Butt welding fittings to BS 1965:Part 1 with flanges to BS 4504.	Long cycle inspections for external corrosion and joint integrity.	Generally, pipe systems require little or no maintenance if correctly specified. Any maintenance will be in the form of periodic repainting where materials such as cast iron and some types of plastic (to maintain UV protection) are exposed to the elements or are installed in a corrosive setting.
35	Black carbon steel pipework to BS 3601. Fittings: As condenser water main.		
U	Unclassified, i.e. pipework/fittings, not to above standards.		

	GAS PIPEWORK		
YEARS	*DESCRIPTION*	*INSPECTION*	*MAINTENANCE*
35+	Copper pipework to BS EN 1057 (thin wall, previously Table X). Fittings to BS EN 1254:Parts 1 & 2.	Long cycle inspections for external corrosion and joint integrity.	Generally, pipe systems require little or no maintenance if correctly specified. Any maintenance will be in the form of periodic repainting where materials such as cast iron and some types of plastic (to maintain UV protection) are exposed to the elements or are installed in a corrosive setting.
35+	Black mild steel pipework to BS 1387 – heavyweight. Fittings: as cold water.		
35+	Black carbon steel pipework to BS 3601. Fittings: as condenser water.		
35+	Ductile iron pipework to BS EN 969. Fittings: as above.		
35	Flexible corrugated stainless steel semi-rigid pipework to BS 7838.		
U	Unclassified, i.e. pipework/fittings, not to above standards.		

	OIL PIPEWORK		
YEARS	*DESCRIPTION*	*INSPECTION*	*MAINTENANCE*
35+	Black mild steel pipework to BS 1387 – heavyweight. Fittings: as cold water	As above.	As above.
U	Unclassified, i.e. pipework/fittings, not to above standards.		

Adjustment factors

Installed in adverse (but not severely corrosive) environments: –5 years

Not sleeved through walls: –5 years

Note: The above factors are not cumulative: the factor that is the largest should be applied.

Assumptions – Design & Installation

Design and installation of the pipework distribution system to be in strict accordance with manufacturer's instructions and industry good practice (eg CIBSE/BSRIA/HVCA/IOP guidance and relevant MoD standards). If relevant, follow guidance in BS 8313. Pay particular attention to the selection and compatibility of materials in the system.

Adequate provision to be included for the thermal expansion of elements of the system.

Pipework supports in accordance with relevant standards cited above, and with manufacturers' recommendations. Supports to be compatible with pipework materials. CIBSE National Engineering Specification clause 4220 provides guidance tables on support spacings for different pipework materials.

Pipework to be sleeved where it passes through walls or other structural openings.

Adequate measures should be taken to prevent freezing in the system.

The appropriate material should be specified or the appropriate protection and/or protective finishes should be applied to delay corrosion or damage caused by the local environment. Ungalvanized steel pipework should be protected by red lead primer or similar. Appropriate action must be taken to prevent galvanic corrosion between dissimilar metals.

Provide appropriate earth cover to buried pipework. Underground pipework to be protected against corrosion, e.g. by application of waterproof sealing tape.

Pipework buried in screeds to be suitable protected. Provision for thermal movement to be provided where appropriate (e.g. plastic sheathing).

Adequate water treatment (e.g. softening) must be provided to suit the local water quality, especially when higher temperatures are used.

Strainer/filters should be fitted to prevent potential problems with scale and debris build-up.

All soldering and welding fluxes to be applied in accordance with manufacturers' instructions. All soldering and welding fluxes to be removed internally and externally according to manufacturers' instructions.

In areas where dezincification of brass occurs, gunmetal or copper fittings must be used.

Extra care should be taken with the jointing of plastic pipe. Manufacturer's guidance must be followed. Only the manufacturer's specified jointing materials must be used.

Assumptions – Commissioning

Pre-commissioning cleaning and the commissioning of water systems to be carried out in accordance with industry good practice (e.g. BSRIA AG8/91 & AG2/89).

Commissioning to be in strict accordance with manufacturer's instructions and relevant BSRIA/CIBSE/HVCA/IOP/MoD/other guidance/codes.

Key failure modes

External corrosion.

Internal fouling, scaling, corrosion and vibration, leading to tube leakage and cracking.

Damage due to excessively high liquid velocity, temperature and/or pressure.

Reduced flow rate due to partially or fully blocked pipework.

Key durability issues

Corrosion resistance of base materials.

Overall water quality and suitability of water treatment used.

Quality of handling, installation and commissioning.

Quality of protection and/or protective coating on materials to prevent corrosion from the environment in which they are located.

Copper is subject to soft water corrosion in certain conditions. Copper is also subject to attack by demineralised water.

Notes

Pipe fittings and jointing materials acceptable for water bylaw purposes are listed in the Water fittings and materials directory published by WRC.

With regard to ductile iron pipe in contact with potable water, attention is drawn to BS EN 545: 1994 section 4.1.4. Care should be taken in the specification of suitable materials, especially in relation to pipe linings & coverings, to ensure that there is no degradation of water quality.

PRE-INSULATED UNDERGROUND PIPEWORK

YEARS	DESCRIPTION	INSPECTION	MAINTENANCE
35+	Steel cased systems with air-gap, to BS 4508:Part 1, comprising steel service pipe(s) insulated with calcuim silicate or high density mineral wool, supported within a steel casing. Casing coated with epoxy resin and a bitumen or coal tar impregnated reinforced wrap.	Periodic inspection for corrosion, damage, joint/ fixing integrity and for water ingress through casing.	Generally, pipe systems require little or no maintenance if correctly specified and installed.
35+	Steel cased systems without air-gap, to BS 4508:Part 4, comprising steel service pipe(s) insulated with calcuim silicate or high density mineral wool, enclosed in a steel casing. Casing coated with epoxy resin and a bitumen or coal tar impregnated reinforced wrap.	Periodic testing of cathodic protection installation. Provision may be made in the system for periodic or continuous fault monitoring, e.g. of liquid	Any leaks should be repaired in a timely manner in order to minimise damage to insulation and casings. Periodic maintenance of cathodic protection
30	Plastic cased systems without air gap, to BS EN 253, comprising steel service pipe bonded to rigid polyurethane foam insulation and a high density polyethylene casing.	ingress into the casing, or liquid collecting in sumps. Continuous monitoring is recommended because of the rapid damage that can result from	installations.
25	Plastics-in-plastics systems comprising polybutylene or cross-linked polyethylene service pipe to BS 7291 encased in polymeric foam insulation and high density polyethylene casing.	water ingress.	
U	Unclassified, i.e. pre-insulated pipework systems not to above standards.		

Adjustment factors

None.

Assumptions – Design & Installation

Design and installation in accordance with BS 7572, BS 4508 and/or BS EN 253, and with manufacturer's instructions and industry good practice (eg CIBSE/BSRIA/HVCA/IOP guidance and relevant MoD standards).

Metal cased systems to be provided with cathodic protection against corrosion (to BS 7361:Part 1) unless deemed unnecessary by a local soil sample.

Dissimilar metals to be isolated to prevent galvanic corrosion. Appropriate soil testing should be undertaken to establish the suitability of the pipework casing material.

Pipework to be suitably protected against surcharge and mechanical damage. Minimum 500mm cover to pipework; minimum 600mm cover under roadways.

Metal casings to have suitable corrosion treatment to internal face prior to assembly.

Metal cased systems with air gap to be provided with internal supports at such intervals that the safe working stress on the casing is not exceeded.

Preformed pipework sections must be handled, transported and stored with caution to prevent damage.

Pipework bedding and backfill to be of appropriate materials and suitably compacted. Materials should be approved as suitable by the manufacturer.

Adequate access chambers to be provided for future maintenance and inspection.

Metal cased systems with air gap are to be laid to a fall of not less than 1 in 500 to facilitate drainage of the casing.

In systems without air gap, any corrosion protection to metal service pipe must be chemically compatible with the insulation. Where the metal service pipe has no anti-corrosion coating, the insulation must provide such protection.

Adequate provision to be included for the thermal expansion of elements of the system.

Adequate water treatment (e.g. softening) must be provided to suit the local water quality, especially when higher temperatures are used.

Strainer/filters should be fitted to prevent potential problems with scale and debris build-up.

Any pipe passing though a wall should be adequately sleeved to allow for movement.

Fitting assemblies, valve assemblies and joint assemblies for pre-insulated bonded pipe systems to be in accordance with BS EN 448, BS EN 488 and BS EN 489 respectively.

Assumptions – Commissioning

Pre-commissioning cleaning and the commissioning of water systems to be carried out in accordance with industry good practice (e.g. BSRIA AG8/91 & AG2/89).

Commissioning to be in strict accordance with manufacturer's instructions and relevant BSRIA/CIBSE/HVCA/IOP/MoD/other guidance/codes.

Particular care should be taken to ensure that the completed installation is free from all foreign matter such as soil, backfill material, surface rust and mill scale. BS 7572 (clause 14) provides guidance on cleaning methods.

Key failure modes

Corrosion of metal casing/service pipe due to ingress of ground water or leaking of service pipe.

Mechanical failure due to overstressing (thermal or pressure stresses, ground loadings), or corrosion.

Degradation and loss of strength of polyurethane foam insulation. (Note that this degradation process is both time and temperature dependent. BS 7572 suggests an average life of 30 years.)

Internal fouling, scaling, corrosion and vibration, leading to tube leakage and cracking.

Damage to service pipe due to excessively high liquid velocity, temperature and/or pressure.

Reduced flow rate due to partially or fully blocked pipework.

Key durability issues

Structural strength of foamed thermal insulation.

Quality of handling, installation and commissioning. BS 7572 cites effective jointing of insulation and casing as the most critical aspect of pre-insulated systems.

Thickness and structural strength of casing.

Provision for thermal movement.

Corrosion resistance of base materials. Protective measures taken, i.e. surface coatings, cathodic protection.

Overall water quality and adequacy of water treatment used.

Particular care is required at building entries, which are a potential source of water ingress to the pipe system.

Effective control of the heating system, water quality, operating temperature and pressure is essential in order to prevent premature failure of the system.

Notes

The risk of damage due to the high compressive stresses that can arise if thermal expansion is restrained by surrounding soil can be reduced by pre-heating or pre-stressing of the pipe prior to backfilling. Further guidance is provided in BS 7572.

VALVES FOR WATER SUPPLY SYSTEMS

Scope

This section provides data on common valve types used for water supply systems in non-domestic buildings. This work only covers manually operated valves for connection to copper or galvanised iron/steel piped distribution systems used for the supply of water of potable quality or for ablution. Valves for use with heating or mediums other than water (gases, hydrocarbons, chemicals, foods etc) and air are excluded from this study.

The following component sub-types are included within this section:

Standards cited

BS 143 & 1256:1986	Specification for malleable cast iron and cast copper alloy threaded pipe fittings.
BS 1010: Part 2:1973	Specification for draw-off taps & stopvalves for water services (screwdown pattern) Draw off taps and above-ground stopvalves
BS 1212	Float operated valves (various parts).
BS 1560: Part 3:1989	Circular flanges for pipes, valves and fittings (Class designated) Steel, cast iron and copper alloy flanges.
BS 2456:1990	Specification for floats (plastics) for float operated valves for cold water services
BS 4504: Part 3:1989	Circular flanges for pipes, valves and fittings (PN designated). Steel, cast iron and copper alloy flanges.
BS 5150:1990	Specification for cast iron gate valves.
BS 5151:1974	Specification for cast iron gate (parallel slide) valves for general purposes.
BS 5152:1974	Specification for cast iron globe and globe stop and check valves for general purposes.
BS 5153:1974	Specification for cast iron check valves for general purposes.
BS 5154:1991	Specification for copper alloy globe, globe stop and check, check and gate valves.
BS 5157:1989	Specification for steel gate (parallel slide) valves.
BS 5159:1974	Specification for cast iron and carbon steel ball valves for general purposes.
BS 5160:1989	Specification for steel globe valves, globe stop and check valves and lift type check valves.
BS 5163:1986	Specification for predominantly key operated cast iron gate valves for waterworks.
BS 5433:1976	Specification for underground stopvalves for water services
BS 6282:1982	Devices with moving parts for the prevention of contamination of water by backflow.
BS 6683:1985	Guide to installation and use of valves.
BS 6700:1997	Specification for the design, installation, testing and maintenance of services supplying water for domestic use within buildings and their curtilages.
BS 6920: Part 1:1996	Suitability of non-metallic products for use in contact with water intended for human consumption with regard to their effect on the quality of water. Specification.
BS 7438:1991	Specification for steel and copper alloy wafer check valves, single disk, spring-loaded type.
BS 8313:1997	Code of practice for accommodation of building services in ducts.
BS EN 1057:1996	Copper and copper alloys. Seamless, round copper tubes for water and gas in sanitary and heating applications

Other references/information sources

ASHRAE Handbooks	HVAC systems and equipment (1996), SI edition, chapter 41, Valves.
BSRIA Application Guide: AG8/91	Pre-commission cleaning of water systems (1991)
BSRIA Application Guide: AG2/89	The commissioning of water systems in buildings (1989)
CIBSE Commissioning code W:1994	Water distribution systems.
HVCA	Standard Maintenance Specification for Mechanical Services in Buildings. Volumes IV – ancillaries, plumbing and sewerage.
PSA Standard 4	Heating, hot and cold water installations for dwellings – Section 7 – Valves, Taps and Cocks.
Institute of Plumbing (IOP)	The Plumbing Engineering Services Design Guide.
WRc Water Fittings & Materials Directory	Approved products (1999)

VALVES FOR WATER SUPPLY SYSTEMS

GLOBE VALVES

YEARS	DESCRIPTION	INSPECTION	MAINTENANCE
20	Copper alloy globe valves to BS 5154.	*Annual:* inspection, check operation. Periodic checking of flow rates and resetting where required.	*Annual:* Perform full operation cycle. Clean and lubricate mechanism andbody as required.
20	Cast iron globe valves to BS 5152.		
20	Steel globe valves to BS 5160.		
20	Copper alloy screw-down stop valves for installation on underground water service pipes to BS 5433.		Copper alloy screw-down valves: Remove debris from access chamber and ensure cover is operational end effective
U	Unclassified, i.e. globe valve, not to relevant standards.		

GATE VALVES

YEARS	DESCRIPTION	INSPECTION	MAINTENANCE
20	Copper alloy gate valves to BS 5154.	*Three monthly:* inspection, check operation. Periodic inspection of stem packing gland integrity.	*General:* Check gland daily after commissioning for 1 week.
20	Cast iron gate valves to BS 5151.		*Annual:* Perform full operation cycle and check daily for 1 week following this operation. Clean and lubricate mechanism and body as required.
20	Cast iron gate valves to BS 5150.		
20	Cast iron gate valves to BS 5163.		
20	Steel gate valves to BS 5157.		Where installed externally, the frequency of lubrication cycles may need to be increased.
U	Unclassified, i.e. gate valve, not to relevant standards		

CHECK VALVES

YEARS	DESCRIPTION	INSPECTION	MAINTENANCE
15	Copper or copper alloy valves to BS 5154.	*General:* Inspection as directed by manufacturer. Check operation as directed by manufacturer	*General:* As directed by manufacturer.
15	Cast iron check valves to BS 5152 and BS 5153.		
15	Steel check valves to BS 5160.		
U	Unclassified, i.e. check valves, not to relevant standards.		

BALL VALVES

YEARS	DESCRIPTION	INSPECTION	MAINTENANCE
10	Cast iron and carbon steel ball valves to BS 5159.	*Annual:* check operation.	*Annual:* Perform full operation cycle. Clean and lubricate mechanism and body as required.
10	Ball valves of any material, used in accordance with WRc listing approval.		
U	Unclassified, i.e. ball valves, not to relevant standards.		

FLOAT VALVES

YEARS	DESCRIPTION	INSPECTION	MAINTENANCE
15	Piston type float operated valve with copper alloy body to BS 1212: Part 1: 1990.	*Three monthly:* inspection, check operation. Periodic inspection of shut-off level in cistern and seal effectiveness. Check float for leaks and fixing.	*General:* maintenance as required. Replace disk washer or diaphragm as necessary. Adjust shut-off level. Remove scale build up.
10	Diaphragm type float operated valve with copper alloy body to BS 1212: Part 2: 1990.		
U	Unclassified i.e. valves not manufactured to the above standards		

Adjustment factors

Manufacturing process not quality assured to BS EN ISO 9001: –5 years

Installed in adverse (but not severely corrosive) environments: –5 years

Longer operating duration or high frequency of use will reduce the life of the component.

Note: The above factors are not cumulative: the factor that is the largest should be applied.

Assumptions – Design & Installation

Installation to be in strict accordance with manufacturer's instructions and industry good practice (e.g. CIBSE/ASHRAE guidance and relevant MoD standards) and following guidance in BS 6683, BS 6700 and BS 8313.

Valve life is highly dependent on correct specification related to the operating conditions. The expected operating conditions should be clearly specified, i.e. pressure and temperature, flow characteristics and frequency of operation.

Where services systems are altered, enlarged or reduced, then assessment of the new conditions should be made and valves changed/adapted accordingly.

The component materials should be selected to avoid bimetallic corrosion. Consideration should also be given to the effect of the operating environment, fire hazards, thermal shock and pipe stresses.

Normal duplex brass valves are not suitable for use in areas where dezincification occurs (generally 'soft' water areas). Water supply quality should be checked with the local water supplier as supplies may be transported from other areas.

DZR (Dezincification resistant) brass valves should be used in all underground applications in accordance with the New Water Regulations 1998, and where in contact with ground water and marine environments.

Valves should be protected from pollutants and aggressive environments (heat, water, dust, freezing, solvents etc.) during storage with end caps and body covers where necessary.

The obturator should be in the position indicated in the product standard if appropriate, or closed to protect the seat and disk.

Assumptions – Commissioning

Pre-commissioning cleaning of water systems to be carried out in accordance with industry good practice (e.g. BSRIA AG2/89 & AG8/91).

Commissioning to be in strict accordance with manufacturer's instructions and relevant CIBSE/ASHRAE/MoD/other guidance/codes.

Valves (particularly gate type) operating in full bore position for long periods should be left with 1/2 turn to the stop to allow for freeing with reverse movement in case the gate/spindle becomes jammed.

Key failure modes

Globe valves: incorrect application (e.g. due to specification error).

Gate valves: gland leakage, mechanical failure (e.g. due to design error, material failure, maintenance error); mechanical issues).

Check valves: failure to open (e.g. due to incorrect installation in flow direction); chemical actions (i.e. lime build-up, dezincification and oxidation).

Ball valves: operation problems (control handle /devise design error); packing gland problems (incorrect installation location).

Float valves: diaphragm failure (e.g. due to excessive wear and/or water condition causing deterioration); disc failure (e.g. due to incorrect specification, excessive wear); float failure (manufacturing error, tank overfill, fatigue failure); gland problems (e.g. design error, material failure, maintenance error).

Internal fouling, scaling, corrosion (all water-related) and vibration, leading to damage.

Erosion of seats due to excessively high water velocity and/or inadequate water treatment.

Valve seats and disks typically suffer the highest wear.

Key durability issues

Standard of manufacture and quality of materials used for the body.

Appropriate specification of the obturator (seat, disk, gate, seal etc) in relation to the operating conditions.

Quality of handling, installation and commissioning.

Corrosion resistance of materials; applied corrosion protection.

Overall water quality and type of water treatment used.

Hours/cycles of operation.

Valve failure is often caused by inadequate specification.

Notes

The Water Research Council (WRc) provide third party assessment under the Water Regulations Advisory Scheme. It produces a list of approved fittings in the Water Fittings and Materials Directory.

Valves for use with potable water supplies from the Water Utility companies must be listed in the Water Fittings and Materials Directory.

Due to the complex performance requirements of valves or their inclusion within or for attachment to other plant, it is common for units to be specified from overseas sources. Specification, recognition and acceptance of foreign standards may therefore be necessary.

CIRCULATION PUMPS FOR HVAC SYSTEMS

Scope

This section provides data on the most common water circulating pump types used in HVAC systems, i.e. centrifugal pumps. It includes direct drive, close coupled and indirect drive pumps. Other, less commonly used types of pumps such as reciprocating pumps and turbine driven pumps are excluded from this section. Submersible drainage/sewerage and sump pumps are also beyond the scope of this study.

Both glanded and glandless pumps are considered within this section, as are different impeller types.

The following component sub-type is included within this section:

Standards cited

BS 970 Series (various parts)	Specification for wrought steels for mechanical and allied engineering purposes
BS 1394 Part 2:1987	Stationary circulation pumps for heating and hot water service systems. Specification for physical and performance requirements.
BS 1400:1985	Specification for copper alloy ingots and copper alloy and high conductivity copper castings.
BS 3100:1991	Specification for steel castings for general engineering purposes
BS 3790:1995	Specification for endless wedge belt drives and endless V-belt drives.
BS 5257:1975	Specification for horizontal end-suction centrifugal pumps (16 bar)
BS 5316 Series (various parts)	Specification for acceptance tests for centrifugal, mixed flow and axial pumps
BS 6861 Part 1:1987	Mechanical vibration. Balance quality requirements of rigid rotors. Methods of determination of permissible residual unbalance
BS EN 1561:1997	Founding. Malleable cast irons
BS EN 25199: 1992	Technical specifications for centrifugal pumps. Class 2
BS EN 60355 Part 2–51: 1997	Stationary circulation pumps for heating and hot water service systems. Safety requirements.
BS EN ISO 9905:1998	Technical specifications for centrifugal pumps. Class 1
BS EN ISO 9908: 1998	Technical specifications for centrifugal pumps. Class 3

Other references/information sources

ASHRAE Handbooks:	HVAC applications (1995); HVAC systems and equipment (1996); Fundamentals (1997)
BS 4082:Parts 1 & 2:1969	Specification for external dimensions for vertical in-line centrifugal pumps (I & U types)
BSRIA Application Guide:	AG8/91 – Pre-commission cleaning of water systems (1991)
BSRIA Application Guide:	AG2/89 – The commissioning of water systems in buildings (1989)
BSRIA Technical manual:	TM1/88 – Commissioning HVAC systems – Division of responsibilities (1988)
CIBSE Guide B:1986	Installation and equipment data – B1 Heating B2/B3 Ventilation and air conditioning B4 Water Service Systems (1986)

Defence Estates Specification 036: Heating, hot and cold water, steam and gas installations for buildings. (Formerly PSA Standard Specification (M&E) No.3).

Electricity at Work Regulations 1989

IEE – Code of practice for in-service inspection and testing of electrical equipment

HVCA	Standard Maintenance Specification for Mechanical Services in Buildings – Volumes 1–5 (1990–92)
Institute of Plumbing (IOP)	The Plumbing Engineering Services Design Guide

CENTRIFUGAL PUMPS

YEARS	DESCRIPTION	INSPECTION	MAINTENANCE
25	Axially split, 'double volute' casing glanded pumps with stainless steel pump shaft to BS 970 and double entry stainless steel impeller to BS 3100. Cast iron, bronze or gunmetal casing.	Inspection in accordance with manufacturer's requirements and HVCA specification, to include (typically): **Weekly:** examine for excessive seal leakage, noise, overheating and vibration, check level/condition of lubricant. **Six monthly:** check lubrication, bearings, impeller, drives and drive couplings (condition and alignment), pulleys, anti-vibration mounts. Inspect glands, strainers, flexible pipe couplings, valves, pressure switches and level controls. Check belts, guards, holding down bolts. **Annually:** inspect electrical connections, motor vent slots (six monthly if warranted by local environment), casings, drive guards.	Maintenance in accordance with manufacturer's requirements and HVCA specification, to include (typically): **Weekly:** change-over of duty/standby pumps **Six monthly:** lubricate pump and motor bearings (unless sealed for life), adjust drives and drive couplings, belts, repack glands, clean/replace strainers (if necessary), replace mechanical seals if leakage unacceptable. **Annually:** clean electrical connections, motor vent slots (six monthly if warranted by local environment), casings. Adjust drive guards (if necessary). Periodic dismantling and overhaul, and replacement of damaged/worn components, gaskets and seals. Replacement of bearings as necessary, depending upon in-service conditions and hours operation. Replacement of motors – see separate section on HVAC motors.
20	Glanded pumps with stainless steel pump shaft to BS 970 and stainless steel impeller to BS 3100. Tungsten carbide seal ring and ceramic seat with EPDM (ethylene propylene) seat insert. Cast iron, bronze or gunmetal casing.		
20	Glanded pumps with stainless steel pump shaft to BS 970 and stainless steel impeller to BS 3100. Carbon seal ring and ceramic seat with EPDM (ethylene propylene) seat insert. Cast iron, bronze or gunmetal casing.		
10	Glanded pumps with stainless steel pump shaft to BS 970 and bronze impeller to BS 1400:LG2. Carbon seal ring and ceramic seat with EPDM (ethylene propylene) seat insert. Cast iron, bronze or gunmetal casing.		
10	Glanded pumps with stainless steel pump shaft to BS 970 and cast iron impeller. Carbon seal ring and ceramic seat with fibre sealing ring. Cast iron, bronze or gunmetal casing.		
10	Glanded pumps with stainless steel pump shaft to BS 970 and Noryl impeller. Carbon seal ring and ceramic seat with fibre sealing ring. Cast iron, bronze or gunmetal casing.		
5	Glandless pumps with stainless steel shaft and sleeve bearing with stainless steel, cast iron, bronze or thermoplastic resin impeller. Standard body seal. Cast iron or cast steel casing.		
U	Unclassified i.e. pump construction unknown, or not to relevant BS, BS EN or BS EN ISO standards.		

Adjustment factors

Installed in adverse (but not severely corrosive) environments: –5 years.

The above lives assume 5 day a week, 12 hour per day operation. Longer operation will reduce pump life. Continuous operation (24 hours / 7 day week): –5 years.

Dual pump system with automatic or managed changeover: +5 years.

Pumped fluids de-oxygenated and free of suspended solids, chemical or other contamination: +5 years.

Note: the above factors are not cumulative. The factor which is the largest should be applied.

Assumptions – Design & Installation

Design, selection and installation of the pump(s) to be in strict accordance with manufacturer's instructions and industry good practice (eg CIBSE/BSRIA /HVCA/ IOP guidance and relevant MoD standards). Pumps to comply with the requirements of BS EN 60355–2–51, BS 4982:Parts 1 & 2 and BS 5257 (as applicable).

Pumps to comply with BS EN ISO 9905 (Class I pumps), BS EN 25199 (Class II pumps), or BS EN ISO 9908 (Class III pumps). Pumps of rated input not exceeding 300W to comply with BS 1394 (see HAPM Component Life Manual P. 6.20). The appropriate class of pump must be selected for the intended use, with regard to required reliability, operating and environmental conditions (with suitable protective finish, as recommended by manufacturers). Pumps should be 'type' tested in accordance with the requirements of BS 5316, and should be selected to provide the correct fluid flow rate.

Adequate provision to be included for the thermal expansion of any elements of the pump and/or the HVAC system. Adequate measures must be taken to prevent potential frost damage.

Pump casings/enclosures are to be of appropriate materials for the intended use/environment. Appropriate corrosion protection to be provided to ferrous materials. Cast iron casings to BS EN 1561. Bronze casings to BS 1400. Appropriate pump bases, vibration isolation, noise insulation and pipework connections to be provided. Direct drive pumps and motors should be mounted on a common bed plate.

Motors to be fan cooled, with thermal overload (built in or separate).

Belt drives to BS 3790.

Pipework to pump must be independently supported and should not transmit any strain to the pump.

Gland packing material and construction to be suitable for the operating conditions.

A strainer/filter must be fitted in the supply line to prevent problems with scale/debris build up.

Pumps and their drives are to be segregated such that failure of the pump seals does not result in damage to drive motors.

Pumps to be installed as directed by the manufacturer, i.e. vertically or horizontally in the pipework. It is essential that the pump does not fall below the horizontal/vertical plane, even by a few degrees, as this causes premature wear. Horizontal pumps must never be installed with the shaft in a vertical plane, as this may lead to dry running and failure.

Pumps should be installed clear of heat sources with pipework properly supported on both sides of the pump.

Pumps to be fitted with isolating valves, a drain plug and an air release cock (if required).

Assumptions – Commissioning

Pre-commissioning cleaning of water systems to be carried out in accordance with industry good practice (e.g. BSRIA AG2/89 & AG8/91).

Commissioning to be in strict accordance with manufacturer's instructions and relevant BSRIA/CIBSE/HVCA/MoD/other guidance/codes.

To prevent early seal failure, it is essential that the pump chamber is filled with liquid before starting the motor.

Key failure modes

Bearing failure, e.g. due to wear, increased temperature, inadequate lubrication and/or the ingress of dirt/water. BS EN ISO 9905 requires a bearing life of at least three years, i.e. 25000 hours continual use. Accurate alignment of impeller and shaft is essential to bearing life.

Mechanical seal failure. Note that dry running will lead to early seal failure.

Cavitation erosion, e.g. due to abnormally high/low flow rates.

Corrosion: erosion corrosion (impeller and wearing rings are particularly susceptible), corrosion fatigue, graphitic corrosion, intergranular corrosion, 'wet end' corrosion of shaft.

Operational faults: fracture due to thermal shock (i.e. rapid changes in temperature of liquid within the pump); exceeding maximum 'head' limits (i.e. low flow rates due to system nearly shut and/or pump oversized); overheating, overload (e.g. due to operation at abnormally high/low flow rates); dry running; abrasive wear of impeller (i.e. due to solids and debris suspended in the water).

Installation faults: incorrect rotational direction; inadequate pipework support.

Key durability issues

Corrosion and wear resistance of pump materials.

Quality/frequency of maintenance, particularly bearing lubrication.

Operation at low flow rates can lead to overheating and reduced motor and bearing life.

Accurate levelling and balancing of the complete pump/drive unit is essential to prevent premature failure of bearings, drive shaft and motor.

Abrasive-wear resistance of pump materials (best is 400-series stainless steel).

Cooling coils may be required to maintain the bearing oil temperature within acceptable limits as defined in the relevant British Standards.

Notes

The above descriptions cover pumps of rated input above 300W only. These may be in-line or end suction circulators, or secondary water supply circulators.

Pumps can be damaged in transit and can be subject to corrosion after factory tests and before installation. Packaging and carriage for centrifugal pumps is covered by BS EN 25199 and BS EN ISO 9905.

Secondary water supply pumps should be of a type approved by the Water Research Council (WRC), in order to comply with the Water Regulations

The recommended maintenance frequencies should be increased in arduous or dusty environments.

Axially split 'double volute' casing pumps with double entry impeller are designed to balance the hydraulic radial loads, thereby minimising shaft deflection and bearing loadings.

For guidance on motor specification and performance, see separate section on HVAC motors.

WATER TREATMENT – BASE EXCHANGE WATER SOFTENERS

Scope

This study provides data on base exchange water softeners for use in commercial and light industrial applications. It does not include systems used for heavy industrial/ manufacturing processes or specialised industry types such as pharmaceutical, brewing or semi-conductor manufacture.

Magnetic and electric-charge scale inhibitors are also excluded from the scope of this study.

The following component sub-type is included within this section:

Component Sub-type	Page
Base exchange water softeners	23

Standards cited

British Effluent and Water Association (BEWA) Code Of Practice 01.85:1985. Code of practice for salt regenerated ion exchange water softeners for direct connection to the mains water supply.

BS 2486:1997	Recommendations for treatment of water for steam boilers and water heaters.
BS 6700:1997	Specification for design, installation testing and maintenance of services supplying water for domestic use within buildings and their curtilages
BS 6920	Suitability of non-metallic products for use in contact with water intended for human consumption with regard to their effect on the quality of the water
Part 1:1996	Specification
Part 2: 1996	Methods of test

New Water Regulations: July 1998 (replaced Water Byelaws).

Other references/information sources

WRc	Water fittings and materials directory (published six monthly).
BW Type Performance Standard: P.19.93:1993	Standard method for testing the performance of salt regenerated ion exchange water softeners for direct connection to the mains water supply.
BSRIA Application Guide:1989: AG 2/89	Commissioning of water systems in buildings
BSRIA Application Guide:1991: AG8/91	Pre-commission cleaning of water systems
BSRIA FMS 4/99: 1999	Guidance and the standard specification for water risk assessment
CIBSE Guide G: 1999: Section 5: Section 6:	Corrosion and corrosion protection Water treatment.
CIBSE Commissioning Code B: 1975 Code W:	Boiler plant, Water distribution system.
HVCA	Standard Maintenance Specification for Mechanical Services in Buildings.
Institute of Plumbing :	Plumbing engineering services design guide
PSA/DEO MEEG 1/03:1987 MEEG 1/05:1989	Water treatment Hot and cold water services.
PSA/DEO TICE 140: 1979	Treatment of potable water
PSA/DEO MEEG 4/03: 1989	Water treatment

Water Supply (Water Quality) Regulations: 1989 (as amended by the Water Supply (Water Quality) (Amendment) Regulations 1991.

WATER TREATMENT *(continued)*

BASE EXCHANGE WATER SOFTENERS

YEARS	DESCRIPTION	INSPECTION	MAINTENANCE
20	Water softeners comprising separate and replaceable resin tank, brine tank and control valves. Design and installation in accordance with BEWA Code of Practice 01.85. Control valve to conform to a 100,000 cycle, high pressure test.	Water treatment equipment should be inspected by 'competent persons' in accordance with the Water Regulations 1999.	Maintain in accordance with manufacturer's requirements and the Water Regulations 1999, to include (typically):
15	Cabinet type water softeners comprising resin tank, brine tank and control valves, contained within a single enclosure. Design and installation in accordance with BEWA Code of Practice 01.85. Control valve to conform to a 100,000 cycle, high pressure test.	*Weekly:* examine for any seal leakage, noise or vibration; test water using hardness tablets. *Monthly:* check water quality, check salt level in brine tank monthly (or daily/ weekly, depending on regeneration frequency.	*Bi-annually:* clean brine tank, use resin cleaner (if required), strip softener control valve and replace seals. *Periodic:* Replenish salts as necessary.
U	Unclassified i.e. base exchange water softeners not to the above standards.	*Bi-annually:* inspect brine tank, check for corrosion/scale build up, check electrical connections. *Annually:* carry out a visual resin sample check, followed by a chemical check if deemed necessary. Equipment should also be subjected to visual inspection during commissioning (and then at least annually) of the main components i.e. associated pipework and filters, resin and brine tanks, distributor, control valves and regeneration controls. Carry out resin sampling if treated water is not of the standard required or damage to resin by iron or chlorine is suspected.	The system is likely to require periodic back-washing to remove deposit build up. Periodic washing with disinfectant may also be necessary to remove any microbiological build-up. The maintenance and inspection frequencies may need to be increased if the incoming water is excessively hard or if the water supplied to other equipment is required to be of a particularly exacting standard. Good operational practice will include the keeping a log of pressures, brine and hardness levels. Replacement of resin beads may be necessary during the life of the softener, particularly if exposed to oxidising agents such as chlorine.

Adjustment factors

The lives assigned above assume a 5 day per week, 12 hour per day operation. Longer operating durations will reduce the service life of the component.

Assumptions – Design & Installation

Design and installation to be in strict accordance with manufacturer's instructions and industry good practice (e.g. BSRIA / CIBSE /HVCA guidance and relevant MoD standards and following guidance in BS 6700 and BS 6920.

Housings and connections to be of corrosion resistant material or suitably coated to prevent corrosion.

Resin beads must be supplied with a certificate of conformity by the manufacturer of the softener. Resin grade selected to match the specific application and operating conditions.

Type and grade of salt must comply with manufacturer's recommendations.

Brine tank to be of plastic, plastic lined steel, or other corrosion resistant materials.

Adequate provision to be included for the thermal expansion of elements.

Adequate measures to be taken to prevent freezing in the unit.

A strainer/filter is to be fitted in the supply line to prevent failure or reduced performance due to suspended material or sediment.

The equipment must be selected to suit the particular application, i.e. pressure, temperature, water hardness (temporary and permanent), alkalinity and dissolved gases.

Pressure drop through the water softener should not exceed 1.5 bar at maximum flow.

An adequate supply pressure (typically minimum 3.0 bar) is needed to ensure correct operation of the softener control valve and to give the correct brine draw.

Water softeners with inlet/outlet connections of 35mm or greater should be supplied with an anti-hammer check valve and flexible connections.

Sample taps should be provided before and after the water softener.

A double check valve is required before the water softener when installed directly on a cold water mains supply. A valved bypass must also be provided.

Water treatment equipment must be installed clear of heat sources.

It is essential to select the most suitable type of valve control unit for the intended operation. Volumetric and duplex systems (operating in dirty/standby mode) tend to be the most cost-effective to operate.

Water equipment that complies with the Water Regulations (which replace Water Byelaws) is listed in the WRc Water fittings & materials directory.

Assumptions – Commissioning

Commissioning to be in strict accordance with manufacturer's instructions and relevant BSRIA, CIBSE HVCA, MoD or other industry guidance and codes of practice. Commissioning of water systems in general is to be in accordance with BSRIA Application Guide AG 2/89 and CIBSE Commissioning Code W.

Pre-commissioning flushing of water systems to be carried out in accordance with good practice, i.e. BSRIA Application Guide AG 8/91.

Key failure modes

Corrosion of resin/brine tank.

Corrosion of control valves.

Sticking/seizure of solenoid valves.

Fracture of control valves.

Deterioration/loss of performance of resin beads, e.g. due to attack by chlorine, fouling by metals (e.g. iron, manganese) in water, or by algae or bacteria.

Blocking of flow paths due to build up of corrosion products.

Key durability issues

Quality of installation and commissioning.

Corrosion resistance of resin/brine tank.

Hours / volume of operation.

Cabinet models, in which the brine tank and resin tank are located within a single enclosure, are assigned a reduced life due to risk of corrosion of control valve located in close proximity to brine tank, and also difficulty/inability to clean brine tank.

Availability of spare parts, especially valves.

Notes

The Water Research Council carries out testing of water fittings and materials.

Seals may require regular replacement due to wear and tear. They can also be damaged by grit in the incoming water supply.

The effectiveness of the resin beads can be compromised by wear and tear; erosion, exposure to high levels of chlorine, iron, manganese etc.

If the incoming water passes through the resin too quickly, only partial softening will take place.

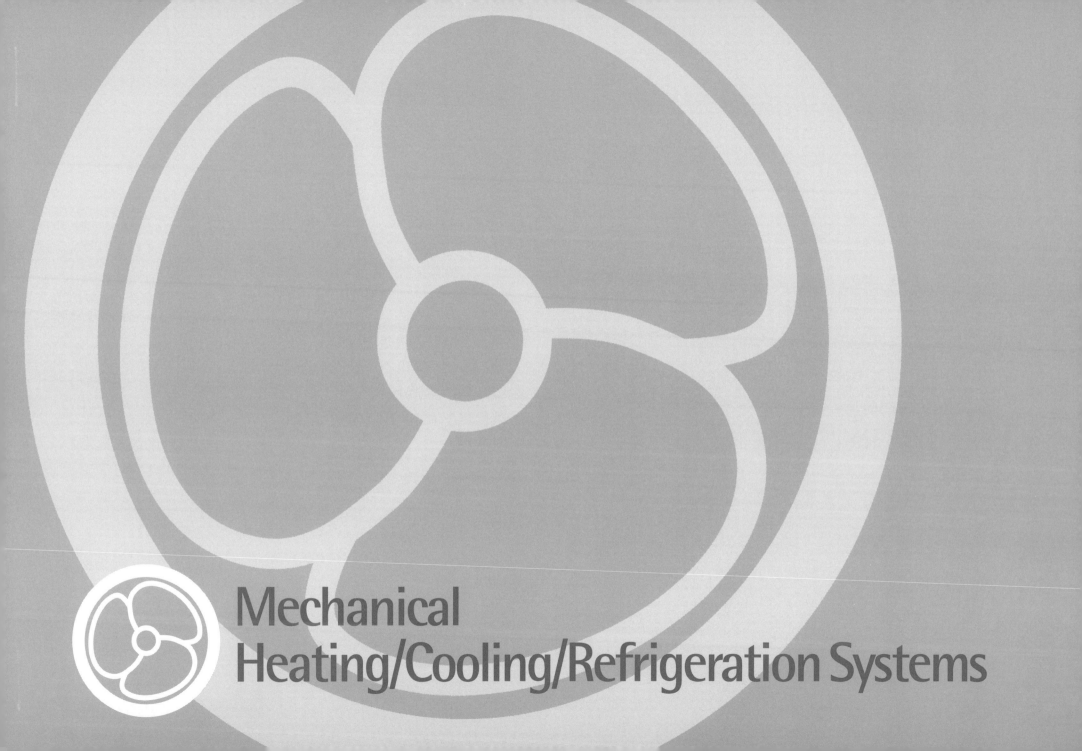

Mechanical
Heating/Cooling/Refrigeration Systems

BOILERS

Scope

This section provides data on boilers (>44 kW rating) commonly used in commercial/light industrial buildings for space and hot water heating purposes. It covers oil and gas fired boilers, but not solid fuel or electrode boilers. Data is provided on the two main construction types, namely cast iron sectional boilers and welded steel boilers.

The following component sub-types are included within this section:

Standards cited

BS 759	Valves, gauges and other safety fittings for application to boilers and to piping installations for and in connection with boilers
Part 1:1984	Specification for valves, mountings and fittings.
BS 779: 1989	Specification for cast iron boilers for central heating and indirect hot water supply (rated output 44 kW and above)
BS 855: 1990	Specification for welded steel boilers for central heating and indirect hot water supply (rated output 44 kW to 3 MW)
BS 2486:1997	Recommendations for treatment of water for steam boilers and water heaters.
BS 2790: 1992	Specification for design and manufacture of shell boilers of welded construction.
BS 3059	Steel boiler and superheater tubes
Part 2:1990	Specification for carbon, alloy and austenitic stainless steel tubes with specified elevated temperature properties.
BS 5885	Automatic gas burners
Part 1: 1988	Specification for burners with input rating 60kW and above.
BS 5978:	Safety and performance of gas-fired hot water boilers (60 kW to 2 MW input)
Part 1: 1989	Specification for general requirements.
Part 2: 1989	Specification for additional requirements for boilers with atmospheric burners
Part 3: 1989	Specification for additional requirements for boilers or induced draught burners.
BS 6644: 1991	Specification for installation of gas-fired hot water boilers of rated inputs between 60 kW and 2 MW (2nd and 3rd family gases).
BS 6798: 1987	Specification for installation of gas-fired hot water boilers of rated input not exceeding 60 kW.

Other references/information sources

BSRIA TN 5/83: 1983	The effect of maintenance on boiler efficiency.
BSRIA TN 2/90: 1990	Investigating boiler failures: A guide for building services engineers.
BSRIA Application Guides: AG2/89	The commissioning of water systems in buildings (1989).
BSRIA Application Guides: AG8/91	Pre-commission cleaning of water systems (1991)
Defence Estates Specification 036:	Heating, hot and cold water, steam and gas installations for buildings. (Formerly PSA Standard Specification (M&E) No.3).
HSE guidance note PM5:1989:	Automatically controlled steam and hot water boilers.

BOILERS *(continued)*

CAST IRON SECTIONAL

YEARS	DESCRIPTION	INSPECTION	MAINTENANCE
25	Cast iron sectional boilers to BS 779, with heat exchanger sections connected with replaceable nipples and held together with tie rods, capable of complete dismantling and reassembly. Burner to be either provided by, or specifically approved by the boiler manufacturer.	Boilers operating at pressures in excess of 0.5 bar above atmospheric pressure are subject to the Pressure Systems and Transportable Gas Containers Regulations 1989, which require statutory safety inspections and maintenance to be carried out by an appropriate 'competent' person. Steam boilers are inspected at least every two years for insurance purposes. Periodic visual inspection of flueways, combustion chambers and connecting flues, for soot and scale. Fire and safety circuits and boiler protection controls should be checked for correct operation at least once a year. Boilers should also be checked regularly for scale and sludge build-up, particularly in hard water areas.	Following inspection, appropriate maintenance should be carried out in accordance with manufacturer's requirements and HVCA Standard Maintenance Specification. It should be noted that maintenance intervals and requirements will vary depending on the usage of the boiler and the specific manufacturer's recommendations. Typical annual maintenance includes: • Testing and maintenance of burners. • Check system water quality and inhibitor type and concentration and top up where necessary. Note that water treatment plant should be maintained to suit the local water conditions, recommendations from any water treatment specialist employed and recommendations of the boiler manufacturer. • Check and clean all flue ways, combustion chamber and connecting flues. Induced/forced draught burners require more regular testing and maintenance (typically 6 monthly) than atmospheric types. Burners may require replacement during the life of the boiler. Typical replacement intervals for atmospheric gas burners are 20–25 years; induced/forced draught and oil burners are 15–20
20	Cast iron sectional boilers to BS 779. Burner to be either provided by, or specifically approved by the boiler manufacturer.		
U	Unclassified cast iron sectional, i.e. cast iron sectional boilers, not to BS 779 or burners not approved by boiler manufacturer.		

WELDED STEEL

YEARS	DESCRIPTION	INSPECTION	MAINTENANCE
20	Welded steel boilers to BS 855. Burner to be either provided by, or specifically approved by the boiler manufacturer.	As above	As above
20	Welded steel shell boilers to BS 2790 (for hot water or steam applications), inspected and pressure tested by an independent third party in accordance with BS 2790 requirements. Burner to be either provided by, or specifically approved by the boiler manufacturer.		
U	Unclassified, i.e. welded steel boilers, not to BS 855 or BS 2790, or burner not approved by manufacturer.		

Adjustment factors

Manufacturing process not quality assured (i.e. to BS EN ISO 9000 series): –5 years.

Lack of adequate water treatment: –5 years (cast iron), –10 years (welded steel).

Sampling of water carried out at least once every year, with dosing of water to adjust water quality (e.g. to neutralise Ph, reduce chlorides etc) and to minimise potentially corrosive chemicals and/or bacteria: +5 years.

Note: The above factors are not cumulative. The factor which is the largest should be applied.

Assumptions – Design & Installation

Design, manufacture and installation of the boiler and the boiler system to be in strict accordance with manufacturer's instructions, industry good practice (e.g. BSRIA/CIBSE/HVCA guidance and relevant MoD standards) and relevant British, European and International Standards.

Gas fired hot water boilers of 60kW to 2MW input are to comply with BS 5978 and are to be installed in accordance with BS 6644. Gas boilers of up to 60kW input are to be installed to BS 6798.

Oil fired boilers should comply with BS 5410:Parts 1 and 2.

The installation should be designed in accordance with the guidance in CP 342:Part 2, the Code of Practice for centralised hot water supply.

Adequate water treatment must be provided to suit the local water quality. This is particularly critical in high temperature applications. Water treatment in accordance with recommendations in BS 2486.

Strainers/filters to be fitted to prevent the build up of scale, debris etc.

Boiler tubes to BS 3059.

The boiler must be selected to suit the particular application, i.e. load, hours of operation and operating conditions (temperature, pressure etc).

Adequate access to be provided for maintenance and inspection.

Valves to be installed in accordance with BS 759. Safety valves to BS 6759.

Steam and hot water boilers to comply with HSE guidance note PM5.

All appliances connected directly or indirectly to the water mains supply should comply with the requirements of the Water Research Council.

Boiler must be equipped with suitable monitoring equipment, including gauges for water flow and return temperatures, dosing pit and draw off cock for sampling of water quality.

Burners must either be supplied by the boiler manufacturer to suit the selected boiler, or to be type matched to the boiler.

Careful control of the return flow temperatures and flow rates, in accordance with manufacturers' directions, is necessary to ensure avoidance of damage caused by acidic condensation, particularly to cast iron or steel elements.

For condensing boilers, a suitable means of collecting and disposing of the condensate must be provided, but copper condensate pipes should be avoided.

Overtightening of tie rods must be avoided.

Assumptions – Commissioning

Pre-commissioning flushing and chemical cleaning of water systems to be carried out in accordance with good practice (e.g. BSRIA AG8/91& AG2/89).

Commissioning and testing to be in strict accordance with manufacturer's instructions and relevant BSRIA /CIBSE /HVCA / MoD/ other guidance / codes before the unit is put into operation.

Burners must be commissioned by a fully trained burner engineer.

Key failure modes

Cracking/fracture due to thermal stress, e.g. as a result of overheating or high temperature differential across the boiler. High temperature is typically caused by poor water flow, poor water distribution in the boiler shell, burner overfiring (both steel and cast iron) or tie rods nor being tightened correctly in cast iron boilers. Build up of corrosion products between boiler sections can result in stress failures.

Thermal stress can also be caused by hot spots due to waterside scale and deposits.

Short term overheating typically caused by operating procedures e.g. water level drop, incorrect operation of pump over-run, large load swing, change in firing pattern etc.

Inadequate flow rates can result in excessive temperature differentials across the boiler, leading to condensation of acidic combustion gases on cooler surfaces, which can result in thinning of boiler shell.

Long term overheating caused by over-firing, deposits, scale build up etc.

Waterside corrosion (caustic or oxygen) caused by poor water treatment and/or boiler protection (both steel and cast iron).

Fireside corrosion e.g. due to condensation inside the boiler which results from a cold return (mainly cast iron).

Manufacturing defects due to poor fabrication and quality control e.g. welding defects.

Failure of seals between boiler sections, e.g. due to distortion/fracture caused by differential expansion.

Burner failures are said to account for some 50% of all boiler failures.

Key durability issues

Standard of manufacture and quality of materials, e.g. thickness, corrosion resistance.

Quality of maintenance and inspection regimes. Maintenance is a critical factor affecting the safety, efficiency and durability of boiler plant. Thorough cleaning, removal of partly burnt fuel, and lime-washing during summer shutdown can help prevent corrosion.

Overall water quality and type of water treatment used. Prevention of scale build up is essential to avoid problems of overheating and reduced water flow. Water treatment programme should be in accordance with BS 2486.

Hours of operation and number of stop/starts.

System design to prevent the average boiler water temperature from falling below 50°C, to ensure that water dew point is not exceeded.

Modes of operation e.g. modes which encourage thermal stress and/or scale build-up.

Steel boilers are particularly susceptible to damage due to build up of sludge or scale.

Notes

Banks of modular boilers can be controlled to fire in various sequences. This means that the lead and lag boilers may be alternated, to average the usage of all boilers. Failure of the lead, or most heavily used boiler is normal in advance of failure of less frequently fired boilers. The above class descriptions are based on the durability of the intermediate boilers, i.e. those which receive medium usage. This can be achieved by a monthly change to the firing sequence to average the usage.

CALORIFIERS

Scope

This section provided data on storage type calorifiers used for heating and hot water supply in non-domestic building types. It includes data on copper, copper lined steel and galvanized steel calorifiers. Glass-lined calorifiers, which are not in common use, are not included in this study. Domestic cylinders are also excluded from this study.

The following component sub-types are included within this section:

Component Sub-type	Page
Copper	35
Copper lined steel	35
Mild steel	35

Standards cited

BS 729: 1971	Specification for hot dip galvanized coatings on iron and steel articles
BS 853	Specification for vessels for use in heating systems
Part 1:1996	Calorifiers & storage vessels for central heating & hot water supply.
Part 2: 1996	Tubular heat exchangers and storage vessels for buildings & industrial services.
BS 2871	Specification for copper & copper alloys. Tubes.
Part 3: 1972	Tubes for heat exchangers.
BS EN 22063:1994	Metallic and other organic coatings. Thermal spraying. Zinc, aluminium and their alloys.

Other references / information sources

ASHRAE Handbooks:	HVAC applications (1995); HVAC systems and equipment (1996); Fundamentals (1997)
BSRIA Application Guide: AG8/91	Pre-commission cleaning of water systems (1991)
BSRIA Application Guide: AG2/89	The commissioning of water systems in buildings (1989)
CIBSE Guide B:1986 Installation and equipment data – B1 Heating, B2/B3 Ventilation and air conditioning (1986)	
CIBSE Commissioning Code W:	Water distribution systems (1994)
DEO/MoD Specification 036: 1997	Heating, hot and cold water, steam and gas installations for buildings
DEO/MoD TB 43/94:1994	Health and safety – work on pressure systems
DEO/MoD SRP 02:1977	Boilers and pressure systems
HSE: 1989 – HSR 30	Guide to the pressure systems and transportable gas containers regulations.

HVCA Standard Maintenance Specification for Mechanical Services in Buildings – Vols 1–5 (1990–92)

CALORIFIERS *(continued)*

COPPER

YEARS	DESCRIPTION	INSPECTION	MAINTENANCE
25	Copper calorifier to BS 853 Grade A, fitted with sacrificial anode. Copper heater tubes to BS 2871. Welding procedure and welder approval to the requirements of BS 853.	Steam calorifiers and water calorifiers operating at pressures in excess of 0.5 bar above atmospheric pressure are subject to the Pressure Systems and Transportable Gas Containers Regulations 1989, which require statutory safety inspections and maintenance to be carried out by an appropriate 'competent' person. Steam calorifiers are inspected at least every two years for insurance purposes.	Following inspection, appropriate maintenance to be carried out in accordance with manufacturer's requirements and HVCA Standard Maintenance Specification. This may include cleaning or replacement of sacrificial anodes, cleaning/removal of lime scale deposits.
25	Copper calorifier to BS 853 Grade B, fitted with sacrificial anode. Copper heater tubes to BS 2871. Welding procedure and welder approval to the requirements of BS 853.		
20	Copper calorifier to BS 853 Grade B, fitted with sacrificial anode. Copper heater tubes to BS 2871. No evidence of welding procedure or welder approval to BS 853.	*Annually:* check operation and safety controls, thermostats, check operation of safety valves, draw off small quantity of water to remove sediment. Check for leaks, bowing/distortion, corrosion. Thickness reading with non-destructive equipment may be required.	
U	Unclassified, i.e. copper calorifier not to to the above standards.	*Biennial:* as annual plus: drain down secondary side and inspect for scale formation and corrosion. Check condition of bursting discs (where fitted). Check sacrificial anodes.	

COPPER LINED STEEL

YEARS	DESCRIPTION	INSPECTION	MAINTENANCE
20	Copper lined steel calorifier to BS 853 Grade A, including anti vacuum valve and sacrificial anode. Heater tubes to BS 2871. Welding procedure and welder approval to the requirements of BS 853.	As above, plus every 24 months check operation of anti-vacuum valve.	As above.
15	Copper lined steel calorifier to BS 853 Grade A, including anti vacuum valve and sacrificial anode. Heater tubes to BS 2871. No evidence of welding procedure or welder approval to BS 853.		
U	Unclassified, ie copper lined steel calorifier not to the above standards.		

MILD STEEL

YEARS	DESCRIPTION	INSPECTION	MAINTENANCE
20	Mild steel calorifier to BS 853 Grade A, with hot dip galvanizing to BS 729 or zinc spray coating to BS EN 22063, to internal surfaces of shell and external surfaces of heater tube. Calorifier fitted with magnesium sacrificial anode. Heater tubes to BS 2871. Welding procedure and welder approval to the requirements of BS 853.	As above.	As above.
15	Mild steel calorifier to BS 853 Grade A, with hot dip galvanizing to BS 729 or zinc spray coating to BS EN 22063, to internal surfaces of shell and external surfaces of heater tube. Calorifier fitted with magnesium sacrificial anode. Heater tubes to BS 2871. Welding procedure/welder not approved to the requirements of BS 853.		
U	Unclassified, i.e. mild steel calorifier not to the above standards, or used in a soft water area.		

Adjustment factors

Manufacturing process not quality assured (i.e. to ISO 9000 series): –5 years.

Calorifiers not fitted with sacrificial anode of a metal suitable for water quality and material compatibility: life limited to 5 years.

Assumptions – Design & Installation

The appropriate calorifier must be selected to suit the application, i.e. pressure, temperature, heating media, water quality.

Installation in strict accordance with manufacturer's instructions.

Design of hot water system to be in accordance with good practice and following guidance in BS 853, with particular regard to the compatibility of materials in the system. Care should be taken to avoid galvanic corrosion between adjacent components. In particular, the use of copper pipework in association with galvanized steel cylinders should be avoided due to the risk of electrolytic action and pitting corrosion. Steel support legs/cradles must be isolated from copper cylinders.

The guidance of the local water authority should be sought as to the suitability of the materials for the local water conditions. The use of galvanized steel cylinders in soft water areas must be avoided.

Adequate water treatment must be provided to suit the local water quality.

Adequate access to be provided for maintenance and inspection.

Copper lined steel calorifiers to be provided with anti-vacuum valve to prevent vacuum within the shell.

Calorifier shell thickness in accordance with BS 853:Part 1.

Assumptions – Commissioning

Pre-commissioning cleaning of the system to remove debris, dust, welding slag etc is particularly important. Commissioning to be carried out in strict accordance with manufacturer's instructions and industry good practice (e.g. BSRIA Application Guides: AG8/91 – Pre-commission cleaning; AG2/89 – The commissioning of water systems in buildings).

Calorifier should be put into service directly after commissioning, and not drained down, which can lead to accelerated corrosion.

Key failure modes

Corrosion, including electrolytic action, pitting corrosion

Leaking, e.g. due to corrosion, cracking, opening of joints or excessive pressure.

Excessive lime scale build-up.

Key durability issues

Standard of manufacture and quality of materials, especially the primary heater i.e. heat exchanger

Quality of welding.

Water quality and type of water treatment used.

Provision of access points for maintenance.

Thickness of copper lining to steel calorifiers.

Provision of an appropriate sacrificial anode reduces the rate of corrosion.

Nature of the heating medium (e.g. hot water, steam, gas or electric heating elements)

Zinc coated steel calorifiers rely on lime scale deposits to protect the zinc layer and are therefore unsuitable for use in soft water areas.

Installation on anti-vibration mountings can help prevent internal and external failure.

MOTORS FOR HVAC SYSTEMS

Scope

This section provides information on the types of electric motor used commonly in HVAC systems. The following motor types are included:

Three-phase squirrel cage induction motors

Single-phase squirrel cage induction motors

Single-phase motors

It does not include information on direct current (DC) motors.

The following component sub-type is included within this section:

Standards cited

BS 2048:1961	Specification for dimensions of fractional horsepower motors
BS 4999:	General requirements for rotating electrical machines (various parts)
BS 5000 (various parts)	Rotating electrical machines of particular types or for particular applications
BS 6107	Rolling bearings: Tolerances (various parts)
BS 6413	Lubricants, industrial oils and related products (Class L).
Part 9:1988	Classification for family X (greases)
BS EN 50081	Electromagnetic compatibility. Generic emission standard (various parts)
BS EN 50082	Electromagnetic compatibility. Generic immunity standard (various parts).
BS EN 60034	Rotating electrical machines (various parts)
BS EN 60529: 1992	Specification for degrees of protection provided by enclosures (IP code)
BS ISO 15:1998	Rolling bearings. Radial bearings. Boundary dimensions. General plan.

Other references/information sources

BS 2757: 1986	Method of determining the thermal classification of electrical insulation
BS 2981: 1979	Methods of test for electric strength of solid insulating materials
BS 3134	Specification for metric tapered roller bearings (various parts)
BS 4480	Specification for plain bearings metric series (various parts)
BS 5512: 1991	Method of calculating dynamic load ratings and rating life of rolling bearings
BS 5856:Part 1:1980 IEC 632–1:1978	Motor starters for voltages above 1 kV ac and 1.2 kV dc
BS ISO 12129/2:1995 ISO 12129:1995	Plain bearings: tolerances on form and position & surface roughness for shaft, flange and thrust collars

HVCA Standard Maintenance Specification for Mechanical Services in Buildings – Volumes 1 to 5.

British Gear Manufacturers' Association.

Rotating Electrical Machines Association.

British Electrical and Allied Manufacturers' Association.

MOTORS FOR HVAC SYSTEMS

YEARS	DESCRIPTION	INSPECTION	MAINTENANCE
15	Totally enclosed fan cooled three-phase squirrel cage induction motor with rolling element regreasable bearings (to BS ISO 15 and BS 6107). Motor protected to IP55 standard or higher and manufactured to BS 2048, BS 4999, BS 5000 and BS EN 60034. Motor constructed from aluminium, cast iron or steel, and fitted with thermistor winding temperature protection. No plastic components.	In accordance with manufacturer's instructions, to include (typically): **Three monthly:** check condition of mountings and bearings **Annually:** check brushes, terminals and connections. Check motor running current. Note: many units are 'sealed-for-life' and are not routinely serviceable. 'Exchange' units may be more usual.	In accordance with manufacturer's instructions and HVCA maintenance specification, to include (typically): **Three monthly:** lubricate bearings where external nipples or lubricators are fitted. Clean away accumulations of lubricant, dust and other debris. **Annually:** replace brushes, clean and test windings. Note: many units are 'sealed-for-life' and are not routinely serviceable. 'Exchange' units may be more usual. Silicon shaft seals fitted to maintain IP55 protection have limited life (depending upon environmental conditions) and should be replaced after approximately 8000 hours of operation.
15	Totally enclosed fan cooled single-phase squirrel cage induction motor, with output not exceeding 6kW, and with rolling element regreasable bearings to BS ISO 15 and BS 6107. Motor protected to IP55 standard or higher and manufactured to BS 2048, BS 4999, BS 5000 and BS EN 60034. Motor constructed from aluminium, cast iron or steel, and fitted with thermistor winding temperature protection. No plastic components.		
10	240V 50Hz single-phase induction motor with output not exceeding 0.37 kW. Motor manufactured to BS 2048, BS 4999 and BS 5000: Part 11. Motor constructed from aluminium, cast iron or steel and fitted with thermistor winding temperature protection.		
5	Motor manufactured to BS 4999 and BS 5000 with protection to IP 22. Motor constructed from aluminium or steel, but having no thermistor winding protection. Motor not selected for specific duty and not operating in a controlled environment.		
U	Unclassified, i.e. motors constructed of unknown materials and/or not to relevant British Standards.		

Adjustment factors

Use of periodic tests and condition monitoring to plot deterioration trend and predict failure: + 5 years.

Hours run recording not available (applies to 10 and 15 year lives only): –5 years.

Sealed-for-life bearings used instead of re-greasable bearings: –5 years (15 year motors only).

Motors used to drive fans (applies to 10 and 15 year lives only): –5 years

Note: The above factors are not cumulative: the factor that is the largest should be applied

Assumptions – Design & Installation

Hours run / time in service to be recorded per motor or via the building management system.

Motor to be rated for and to be compatible with the required duty, to BS EN 60034 Part 1.

Motor to be suitable for intended environment, e.g. appropriate corrosion protection in damp/humid environments.

Motor frame and windings not to be operated at a higher temperature than was planned at design stage. Classes of insulation to be to BS 4999:Part 111. A margin for safety must be provided.

Stator windings to be fitted with phase barriers to reduce or negate effect of excess voltage, to BS 4999 Part 115 or better.

3-phase motors to be designed for 400V ±10%. All motors to be rated for continuous running duty type S1 and be marked as such, to BS 4999:Part 101. Motor operated above 40°C and at high altitude to be de-rated according to manufacturer's requirements.

Built-in thermistor required to detect excessive winding temperature and to trigger power isolation when this occurs.

Appropriate bearings to be selected by manufacturer according to the duty specified and the magnitude of axial or radial thrust.

Motor not to be subjected to excessive starting, i.e. more than 1000 starts per annum.

Motors that are not put into service immediately after leaving the factory must be stored and maintained in strict accordance with manufacturer's instructions.

Lubricant to motor manufacturer's specification. Lithium soap type and non-EP (extreme pressure) grease to BS 6413 and ISO 6743.

Motors are designed for mounting horizontally or vertically but orientation should be specified prior to procurement.

Motors serving belt-driven fans or pumps should be V-belt driven with a minimum of two belts and incorporating pulleys with taper lock fittings.

Installation in strict accordance with manufacturer's instructions and relevant BSRIA/CIBSE/HVCA/ other guidance/codes.

Motor enclosures to have a minimum IP (protection) to BS EN 60529 as indicated, or if continuously exposed to the weather, enclosures to IP54 of the standard. Silicon shaft seals fitted to maintain IP55 protection have limited life (depending upon environmental conditions) and should be replaced after 8000 hours of operation.

Motors should be CE marked to indicate compliance with the Low Voltage Directive.

Electromagnetic compatibility to be to BS EN 50081 Part 2 Emission and BS EN 50082 Part 2 Immunity.

Motor maintenance will be minimal for machines with 'sealed for life' bearings but downtime will have to be allowed for replacement within the component life stated. 'Sealed for life' or double shielded bearings should be replaced before the end of the predicted life of the motor. Double shielded bearings have a typical life of 20,000 to 40,000 hours depending on speed of motor and duty, but refer to manufacturer's information. (For fans, this life is reduced).

Maintenance frequency, as stated in the table, can be reduced or intervals increased, for non-critical (essential and non-essential plant) but only in conjunction with manufacturer's recommendations.

Selection of the type of motor starter employed will depend on the load torque and on the local electricity supply authority's regulations; direct on-line starting may be inadmissible for all but the smallest machines.

Assumptions – Commissioning

Commissioning in strict accordance with manufacturer's instructions and BSRIA Application Guides/CIBSE Commissioning Codes/other appropriate guidance.

Key failure modes

Overheating, e.g. caused by excessively high ambient temperature, unsuitable use, duty exceeds that which was originally specified, incorrect selection. BS 4999:Part 111 states that for every 8–10°C increase in winding temperature, the life of the windings is halved.

Failure due to excessively low speed (i.e. too low to maintain satisfactory lubrication and cooling).

Insulation failure due to excess voltage or voltage spikes between windings, or overheating.

Wear or other problems with drive shaft, belt or sheave.

Defective electrical starting.

Overloading.

Corrosion of casing/winding.

Failure of bearings, e.g. due to wear, misalignment. Bearing failure is particularly common with belt driven plant.

Switching transient damage.

Key durability issues

Life of bearings varies considerably according to the duty and the axial or radial thrust. At the design stage manufacturers to be given full information relating to duty, environment and load. Bearing life is proportional to speed and inversely proportional to the number of pairs of poles. The higher the speed of motor rotation the shorter the bearing life. Bearing life can be affected by different pulley selection for the same duty.

Intensity of use, particularly the number of start/stops.

Standard of initial specification i.e. correct fan for correct application and location.

Standard of manufacture and quality of materials.

Life of motors operated in pairs will be greater, one running and one on standby, approximately twice the component life in the table, provided the duty is alternated periodically.

Life can be prolonged through maintenance and replacement of sub-components linked to hours run recording and condition monitoring.

To prevent overheating problems, a margin of safety is recommended, i.e. provision of a higher class of insulation than that required by standards. E.g. stator to have Class 'F' insulation but for Class 'B' temperature rise.

Ingress protection of IP 55, BS 4999 Part 105, will ensure that water and dust ingress are kept to a practical and economical minimum.

Notes

Sealed for life bearings are often referred to as double shielded. Re-greasable are referred to as single or open shielded. Typically, motors up to 20kW, frame sizes up to 180, will have sealed for life bearings. For sealed bearings, down time must be planned for bearing replacement. Lubrication frequency of bearings varies with motor load in kW, speed and bearing temperature. It is essential that only the lubricant specified by the manufacturer is used.

Most motor maintenance is related to hours of operation rather than pre-defined intervals. Most manufacturers require running hours to be recorded or calculated.

Motor overhaul and bearing replacement are best carried out by the manufacturer or appointed agents. If such work is carried out by local labour then motor and bearing manufacturer's proprietary tools should be used.

PACKAGED CHILLER PLANT – VAPOUR COMPRESSORS

Scope

This component study is concerned with the most common types of compressor unit associated with packaged chiller plant, i.e. reciprocating, centrifugal, screw and absorptive types. It does not include rotary compressors (for small-scale applications) or scroll types.

The following component sub-types are included within this section:

Standards cited

BS 3122	Refrigerant compressors.
Part 1: 1990	Methods of test for performance.
Part 2:	Methods for presentation of performance data.
BS 2871	Specification for copper and copper alloys. Tubes.
Part 3:1972	Tubes for heat exchangers
BS 4434: 1995	Specification for safety and environmental aspects in the design, construction and installation of refrigerating appliances and systems.
BS EN 378–1: 1995	Specification for refrigeration systems and heat pumps. Safety and environmental requirements. Basic requirements.
BS EN 814: 1997	Air conditioners and heat pumps with electrically driven compressors. Cooling mode.
BS EN 12055: 1998	Liquid chilling packages and heat pumps with electrically driven compressors. Cooling mode. Definitions, testing and requirements.
PREN 378	Refrigeration systems and heat pumps. Safety and environmental requirements.

Other references/information sources

ASHRAE Handbook1998: Refrigeration.

BSRIA TM 1/88   Commissioning HVAC systems

Institute of Refrigeration   Safety codes for refrigerating systems utilising group A1 and A2 Refrigerants. 1999.

IOR   Safety codes for refrigerating systems utilising ammonia, 1990.

The Pressure Systems and Transportable Gas Containers Regulations 1989.

CIBSE Guide B14:   Refrigeration and heat rejection

CIBSE Commissioning Code R: Refrigeration

DEO/MoD: Spec 037   Air conditioning, air cooling and mechanical ventilation for buildings

DEO/MoD: MEEG 4/08   Refrigeration systems

HVCA Standard maintenance specifications: Vol 1 – Heating and pipework systems,
Vol 2 – Ventilation and air conditioning.

International Institute of Ammonia – ANSI/IIAR 2–1992, Equipment Design and Installation of Ammonia Mechanical Refrigerating Systems

PACKAGED CHILLER PLANT – VAPOUR COMPRESSORS *(continued)*

RECIPROCATING COMPRESSOR UNITS

YEARS	DESCRIPTION	INSPECTION	MAINTENANCE
20	Reciprocating units manufactured to BS EN 378 and tested to BS EN 12055 and BS 3122. Third party test certificate for performance. ISO 9000 quality assured manufacturing process.	Inspection in accordance with manufacturer's recommendations and IOR safety codes. **Ongoing:** monitoring of water treatment, brine quality and oil use, daily operating conditions through log sheets. Visual inspection every 10,000 run hours.	After inspection, appropriate maintenance in accordance with manufacturer's requirements and IOR/BSRIA/CIBSE/HVCA/other specification, to include (typically):
15	Reciprocating units manufactured to BS EN 378 and tested to BS EN 12055 and BS 3122. ISO 9000 quality assured manufacturing process.		**Six monthly:** Cleaning of fins to remove with dirt and grime build up.
10	Reciprocating units manufactured to BS EN 378 and tested to BS EN 12055 and BS 3122.	**Six monthly:** check suction and discharge pressure, operation of crankcase heater, crankcase oil level and temperature, oil reservoir operating pressures including pressure failure switches. Also check condition and alignment of compressor drive; test safety cut-outs; check for/rectify oil or refrigerant leaks; check refrigerant charge; check control items.	**Annually:** lubricate electric motor. Condenser and lubricant cooler cleaning. Evaporator cleaning on open systems. Calibrating of control items. Tightening of power connections.
U	Unclassified, i.e. reciprocating units not manufactured/tested to the above standards.	**Annually:** check anti-vibration mountings, flexible pipe connections etc. Check tightness of hot gas valve, check all electrical connections including power contacts in starters, dielectric checking of hermetic and open motors and alignment of open motors. Check drive alignment and the applied torsion for holding down bolts of belt driven and direct coupled compressors. Where a belt drive is employed the condition of the belts should be examined and alignment and tension checked. Compressors incorporating liquid refrigeration injection should be examined for correct operation. Service valves incorporating seal caps should be checked for adequate seals and valve stem glands checked for leakage.	

CENTRIFUGAL COMPRESSOR UNITS

YEARS	DESCRIPTION	INSPECTION	MAINTENANCE
20	Centrifugal units manufactured to BS EN 378 and tested to BS EN 12055 and BS 3122. Third party test certificate for performance. ISO 9000 quality assured manufacturing process.	Inspection in accordance with manufacturer's recommendations and IOR safety codes.	After inspection, appropriate maintenance in accordance with manufacturer's requirements and IOR/BSRIA/CIBSE/HVCA/other specification, to include (typically):
15	Centrifugal units manufactured to BS EN 378 and tested to BS EN 12055 and BS 3122. ISO 9000 quality assured manufacturing process.	***Ongoing:*** monitoring of water treatment, brine quality and oil use, daily operating conditions through log sheets. Visual inspection every 10,000 run hours.	
10	Centrifugal units manufactured to BS EN 378 and tested to BS EN 12055 and BS 3122.		***Three monthly:*** tighten holding down bolts and change purge unit filter driers (if required)
U	Unclassified, i.e. centrifugal units not manufactured/tested to the above standards.	***Three monthly:*** check operation, oil pressure and level, refrigerant charge level, operation of purge unit, check holding down bolts.	***Six monthly:*** change oil filter and oil drier return system.
		Six monthly: check control items, oil reservoir operating pressures including pressure failure switches. Check all electrical connections.	***Annually:*** clean purge unit valves and other valves; remove and clean float assembly; change oil & gas filters. Condenser tube cleaning (more frequent if appropriate water treatment is not maintained).
		Annually: sample refrigerant charge, compressor oil; inspect purge unit valves and other valves; check anti-vibration mountings, flexible pipe connections etc.	

SCREW COMPRESSOR UNITS

YEARS	DESCRIPTION	INSPECTION	MAINTENANCE
25	Screw units manufactured to BS EN 378 and tested to BS EN 12055 and BS 3122. Third party test certificate for performance. ISO 9000 quality assured manufacturing process.	Inspection in accordance with manufacturer's recommendations and IOR safety codes.	After inspection, appropriate maintenance in accordance with manufacturer's requirements and IOR/BSRIA/CIBSE/HVCA/other specification, to include (typically):
20	Screw units manufactured to BS EN 378 and tested to BS EN 12055 and BS 3122. ISO 9000 quality assured manufacturing process.	***Ongoing:*** monitoring of water treatment, brine quality and oil use, daily operating conditions through log sheets.	
10	Screw units manufactured to BS EN 378 and tested to BS EN 12055 and BS 3122.		***Six months:*** Water cooled condensers must be cleaned of scale. Change lubricant filter elements.
U	Unclassified, i.e. screw units not manufactured/tested to the above standards.	***Three monthly:*** Leak testing continuously operated process air-cooled units.	***1.5 to 4 years:*** replace hydraulic cylinder seals
		Six monthly: dismantle and inspect non-return valve, check unloading gear, check control items. Flooded units used for year round cooling should be leak tested.	
		Annually: Condition check of refrigerant sample to determine degree of contamination by water, acid or metallic particles. Check all electrical connections, power contacts in starters, dielectric checking of hermetic & open motors and alignment of open motors. Water cooled package should be leak tested.	
		1.5 to 4 years: Inspect shaft seals; ***4 to 6 years:*** Check thrust bearings; ***7 to 10 years:*** Inspect shaft bearings.	

Adjustment factors

Exposure to adverse weather conditions e.g. exposed coastal site: –5 years.

Hermetic compressors: –5 years.

Continuous operation i.e. 24 hours day/ 7 days week: –10 years.

Note: The above factors are not cumulative: the factor which is the largest should be applied.

Assumptions – Design & Installation

All refrigerating systems in which the refrigerant is evaporated and condensed in a closed circuit, including heat pumps and absorption systems, but excluding systems using air as the working fluid should be in accordance with BS 4434:1995 and Institute of Refrigeration (IOR) safety codes for refrigerating systems.

Continuing availability of spare parts should be confirmed.

Installation in strict accordance with manufacturers' instructions and relevant BSRIA/CIBSE/HVCA/other guidance/codes.

Adequate access to be provided for inspection and maintenance.

Motors should be refrigerant cooled, with integral sensing solid state overload protection in each phase.

Refrigerant side to be triple evacuated.

Assumptions – Commissioning

Commissioning in strict accordance with manufacturers' instructions and relevant BSRIA/CIBSE/HVCA/other guidance/codes. Commissioning of new installations should be in accordance with IOR safety code of refrigerating systems utilizing group A1 and A2 refrigerants.

The pressure and temperature ranges in use should correspond to the manufacturer's recommendations.

Rotating components should be statically and dynamically balanced.

Key failure modes

Liquid refrigerant problems i.e. excessive amounts of liquid drawn back into the compressor can damage the valve plates and the cylinder head. Unsuitable type of refrigerant. Presence of liquid refrigerant in the suction lines. Presence of moisture in the refrigerant system.

Compressor overheating i.e. a 25°C increase in discharge temperature could radically shorten the life of a compressor.

Low pressure control failure i.e. operating beyond approved limits can create temperatures in the piston and cylinder area that will reduce the effectiveness of lubrication, causing piston wear which can lead to early failure.

Compressor 'short cycling'.

Cold start and failure of compressor sump heaters.

High discharge pressures i.e. exceeding recommended pressure ranges often or for a long period of time will reduce the life of the compressor

Automatic-reset high pressure failure i.e. could be replaced by manual controls.

Incorrect specification i.e. different requirements from the consultant and the maintenance/facilities engineer.

Key durability issues

Quality of installation and commissioning.

Overall water quality and type of water treatment used.

Hours of operation.

Location e.g. exposure to bad weather and susceptibility to physical damage

Notes

Need for engine maintenance is directly proportional to the running hours per year.

CIBSE Guide section B14 provides guidance on refrigeration plant, including controls.

CIBSE Guide section B7 provides recommendations for water treatment.

PACKAGED CHILLER PLANT (continued)

ABSORPTIVE UNITS

YEARS	DESCRIPTION	INSPECTION	MAINTENANCE
25	Absorptive units manufactured to BS EN 378. Third party test certificate for performance. ISO 9000 quality assured manufacturing process.	**Three monthly:** check refrigerant charge level and operation of purge system. Purge operation should be logged and monitor purge operation trends. Increases purge operation signifies air and moisture ingress. Check pump oil level and motor bearings.	After inspection, appropriate maintenance in accordance with manufacturer's requirements and IOR/BSRIA/CIBSE/HVCA/other specification, to include (typically):
20	Absorptive units manufactured to BS EN 378. ISO 9000 quality assured manufacturing process.		After initial start-up: pump maintenance through cleaning of magnetic strainer every 2 weeks.
10	Absorptive units manufactured to BS EN 378.		
U	Unclassified, i.e. absorptive units not manufactured to the above standards.	Check for leaks, ammonia based systems can be very aggressive. Lithium bromide can be highly corrosive.	**Three monthly:** changing of pump oil and cleaning of magnetic strainer.
		Six monthly: check control items.	**Annually:** Calibrating of control items. Tightening of power connections.
		Annually: sample refrigerant charge, for contamination, pH, corrosion inhibitor level, and performance additive level. Check all electrical connections	**During shutdown periods:** Vacuum pump should be flooded with oil to prevent internal corrosion.
		Three years: shaft seal inspection.	Routinely purging of the unit should be carried out to prevent air leakage.
			Maintained lubrication of pump and motor bearings during start-up after prolonged or seasonal shut down is critical since refrigerant may migrate from the evaporator to the absorption chiller during these periods

Adjustment factors

Exposure to adverse weather conditions i.e. exposed coastal site: –5 years.

Continuous operation i.e. 24 hours day/ 7 days week: –10 years.

Note: The above factors are not cumulative: the factor which is the largest should be applied.

Assumptions – Design & Installation

All refrigerating systems in which the refrigerant is evaporated and condensed in a closed circuit, including heat pumps and absorption systems, but excluding systems using air as the working fluid should be in accordance with BS 4434:1995 and Institute of Refrigeration (IOR) safety codes for refrigerating systems.

Continuing availability of spare parts should be confirmed.

Installation in strict accordance with manufacturers' instructions and relevant BSRIA/CIBSE/HVCA/other guidance/codes.

Adequate access to be provided for inspection and maintenance

Motors should be refrigerant cooled, with integral sensing solid state overload protection in each phase.

Refrigerant side to be triple evacuated.

Additional water storage or float control in the evaporator designed to ensure there is enough water in the machine to protect the refrigerant pump motor from overheating. Designed with hermetic integrity to prevent air leakage into the machine. Anti-crystallization or de-crystallization devices incorporated or heat exchanger bypass lines to facilitate de-crystallization. External connections on pump lubrication system to facilitate auxiliary lubrication necessary after prolonged shut down periods.

Assumptions – Commissioning

Commissioning in strict accordance with manufacturers' instructions and relevant BSRIA/CIBSE/HVCA/other guidance/codes. Commissioning of new installations should be in accordance with IOR safety code of refrigerating systems utilizing group A1 and A2 refrigerants.

The pressure and temperature ranges in use should correspond to the manufacturer's recommendations.

Rotating components should be statically and dynamically balanced.

Key failure modes

Low pressure control failure i.e. operating beyond approved limits can create temperatures in the piston and cylinder area that will reduce the effectiveness of lubrication, causing piston wear which can lead to early failure.

Incorrect specification i.e. different requirements from the consultant and the maintenance/facilities engineer.

Presence of moisture in the refrigerant system.

Presence of large amounts of refrigerant either in the lubricating oil or in the returning suction gas.

Presence of liquid refrigerant in the suction lines.

Freezing where lithium bromide solution becomes too concentrated and form and plug the machine (usually the heat exchanger section). The most frequent causes are air leakage into the machine, low temperature condenser water, electrical power failures

Key durability issues

Standard of manufacture and quality of materials.

Overall water quality and type of water treatment used.

Hours of operation.

Location e.g. exposure to bad weather and susceptibility to physical damage

In absorption machines the compressor is replaced by a heat operated absorber-generator with no moving parts except pumps. Key durability issues relate to keeping the heat transfer surfaces free of scale and sludge, thus maintained water treatment and mechanical sludge removal is of utmost importance.

Significant maintenance issues also relate to extraordinary high vacuum requirements for long periods.

Notes

Need for engine maintenance is directly proportional to the running hours per year.

CIBSE Guide section B14 provides guidance on refrigeration plant, including controls.

CIBSE Guide section B7 provides recommendations for water treatment.

Ventilation/Air Conditioning Systems

HVAC FANS

Scope

This section provides data on the main fan types associated with heating, ventilation and air conditioning systems. It includes centrifugal, mixed flow, axial flow and propeller-type fans. It does not include domestic-type extract fans or fans for specialist applications such as chemical plant or fume cupboards. Bifurcated fans are also excluded from this study.

This study has been divided into two sections: the first section considers directly driven fans with an integral motor; the second section considers belt driven fans.

The following component sub-types are included within this section:

Standards cited

BS 848 (various parts)	Fans for general purposes.
BS 2048:1961	Specification for dimensions of fractional horsepower motors
BS 3790:1995	Specification for endless wedge belt drives and endless v-belt drives.
BS 4999 (various parts)	General requirements for rotating electrical machines
BS 5000 (various parts)	Rotating electrical machines of particular types or for particular applications
BS 5060: 1987	Specification for performance and construction of electric circulating fans and regulators
BS 5285: 1975	Specification. Performance of a.c. electric ventilating fans and regulators for non-industrial use
BS 5720: 1979	Code of practice for mechanical ventilation and air conditioning in buildings
BS 5925: 1991	Code of practice for ventilation principles and designing for natural ventilation
BS 6107 (various parts)	Rolling bearings: Tolerances
BS 8313: 1997	Code of practice for accommodation of building services in ducts
BS EN 60034 (various)	Rotating electrical machines
BS EN 60529:1992	Specification for degree of protection provided by enclosures (IP code)
BS ISO 15:1998	Rolling bearings. Radial bearings. Boundary dimensions. General plan.

Other references/information sources

ASHRAE Handbooks:	HVAC applications (1995); HVAC systems and equipment (1996); Fundamentals (1997)
BSRIA	Commissioning of air systems in buildings:1989

CIBSE Commissioning Code A: 1996 Air distribution systems

CIBSE Guide B:1986 Installation and equipment data – B1 Heating, B2/B3 Ventilation and air conditioning

DEO/TSD Spec 037:1998 Air conditioning, air cooling and mechanical ventilation for buildings

Fan Manufacturer's Association Fan application guide 1981

Fan Manufacturer's Association Fan and ductwork installation guide 1993

HVCA Standard Maintenance Specification for Mechanical Services in Buildings – Volumes 1 to 5

PSA/DEO MEEG 1/06 Air conditioning 1983

PSA/DEO MEEG 1/07 Ventilation 1989

PSA/DEO MEEG 4/06 Air conditioning 1984

PSA/DEO MEEG 4/07 Ventilation 1984

HVAC FANS – DIRECT DRIVEN

CENTRIFUGAL FANS

YEARS	DESCRIPTION	INSPECTION	MAINTENANCE
20	Centrifugal fans manufactured and tested to BS 848, with totally enclosed fan cooled three-phase squirrel cage induction motor with rolling element regreasable bearings (to BS ISO 15 and BS 6107). Motor protected to IP55 standard or higher and manufactured to BS 2048, BS 4999, BS 5000 and BS EN 60034. Motor fitted with thermistor winding temperature protection.	All fans to be inspected in accordance with maunfacturer's requirements, to include (typically): **Six monthly:** inspection of fan/motor bearings (unless 'sealed for life' type), inspection of fan wheels, sheaves and bearing collars, mounting bolts and guide vanes. Check operation of automatic changeover controls. **Annually:** inspection of impeller and spinnings and inspection of motor assembly.	Maintenance in accordance with manufacturer's requirements and HVCA specification, to include (typically): **Six monthly:** lubrication of fan/motor bearings (unless 'sealed for life' type), adjustment of fan wheels, sheaves and bearing collars, mounting bolts and guide vanes. **Annually:** cleaning of impeller and spinnings.
15	Centrifugal fans manufactured and tested to BS 848 with totally enclosed fan cooled single-phase squirrel cage induction motor, with output not exceeding 6kW. Motor protected to IP55 or higher and manufactured to BS 2048, BS 4999, BS 5000 and BS EN 60034. Motor fitted with thermistor winding temperature protection.		
10.	Centrifugal fans manufactured and tested to BS 848, with 240V 50Hz single-phase induction motor with output not exceeding 0.37 kW. Motor manufactured to BS 2048, BS 4999 and BS 5000: Part 11. Motor fitted with thermistor winding temperature protection.		
5	Centrifugal fans manufactured and tested to BS 848, with motor manufactured to BS 4999 and BS 5000 with protection to IP 22. Motor not fitted with thermistor winding temperature protection.		
U	Unclassified i.e. centrifugal fans not manufactured/tested to the above standards.		

MIXED FLOW FANS

YEARS	DESCRIPTION	INSPECTION	MAINTENANCE
20	Mixed flow fans manufactured and tested to BS 848, with totally enclosed fan cooled three-phase squirrel cage induction motor with rolling element regreasable bearings (to BS ISO 15 and BS 6107). Motor protected to IP55 standard or higher and manufactured to BS 2048, BS 4999, BS 5000 and BS EN 60034. Motor fitted with thermistor winding temperature protection.	All fans to be inspected in accordance with maunfacturer's requirements, to include (typically): **Six monthly:** inspection of fan/motor bearings (unless 'sealed for life' type), inspection of fan wheels, sheaves and bearing collars, mounting bolts and guide vanes. Check operation of automatic changeover controls. **Annually:** inspection of impeller and spinnings and inspection of motor assembly.	Maintenance in accordance with manufacturer's requirements and HVCA specification, to include (typically): **Six monthly:** lubrication of fan/motor bearings (unless 'sealed for life' type), adjustment of fan wheels, sheaves and bearing collars, mounting bolts and guide vanes. **Annually:** cleaning of impeller and spinnings.
15	Mixed flow fans manufactured and tested to BS 848 with totally enclosed fan cooled single-phase squirrel cage induction motor, with output not exceeding 6kW. Motor protected to IP55 or higher and manufactured to BS 2048, BS 4999, BS 5000 and BS EN 60034. Motor fitted with thermistor winding temperature protection.		
10	Mixed flow fans manufactured and tested to BS 848, with 240V 50Hz single-phase induction motor with output not exceeding 0.37 kW. Motor manufactured to BS 2048, BS 4999 and BS 5000: Part 11. Motor fitted with thermistor winding temperature protection.		
5	Mixed flow fans manufactured and tested to BS 848, with motor manufactured to BS 4999 and BS 5000 with protection to IP 22. Motor not fitted with thermistor winding temperature protection.		
U	Unclassified i.e. mixed flow fans not manufactured/tested to the above standards.		

Adjustment factors

Dual fan system with automatic or managed changeover: +5 years.

Impellers of insufficient standard for intended use, or not properly balanced: –10 years (or life assessment of 5 years, whichever is greater).

Sealed-for-life bearings used instead of re-greasable bearings: –5 years (20 & 15 year life fans only).

Manufacturing process not quality assured i.e. to ISO 9000 series: –5 years (20, 15 and 10 year life fans only).

The above factors are not cumulative: the factor that is the largest should be applied.

Assumptions – Design & Installation

Fans are to be 'type' tested in accordance with the requirements of BS 848: Parts 1 and 2, and must be selected to deliver the required air volume flow rate. Suitable match to be made between expected usage and related duty and the system type in which the fan is installed.

Fan selection must be appropriate for intended use, i.e. suitability for operation in corrosive environments or in presence of abrasive airborne particles.

Steel casings, frames etc. to be suitably protected against corrosion, e.g. by hot dip galvanizing to BS 729, or aluminum-zinc coating, or appropriate polymeric or stoved enamel coating. Impellers must also be suitably protected. Glass fibre reinforced polymeric impellers are available for use in certain aggressive environments.

Motor enclosures to have a minimum IP (protection) to BS EN 60529 as indicated.

Fan bearings to be suitable for the size, speed, load and discharge angle of the fan.

Check that impeller has adequate clearance and that all local ductwork is properly supported.

Installation in strict accordance with manufacturer's instructions and relevant BSRIA/CIBSE/HVCA/FMA/HEVACA/other guidance/codes.

Foundations and fixing points must suit the dynamic load and frequency of the fan.

Assumptions – Commissioning

Commissioning in strict accordance with manufacturer's instructions and BSRIA Application Guides/CIBSE Commissioning Codes/other appropriate guidance, such as that from the Fan Manufacturers' Association.

Key failure modes

Fans connected to mains supply with wrong electrical characteristics. (Note that some types of fan installed such that they revolve in the wrong direction will still deliver a volume of air, albeit less than the design volume).

Motors not continuously rated.

Corrosion due to inadequate protection/inappropriate selection for intended use/environment.

Fans used/placed in incorrect atmosphere i.e. air handling fans for 'normal' atmospheres used in environment polluted by dust, soot, flour, sawdust etc.

Bearings failure, e.g. due to uneven build up of dust on impellers, causing fan to become out of balance.

Motor failures (N.B. this is a common cause of fan failure).

Impeller failure due to fatigue.

Too much/too little airflow, air surging, excessive noise and vibration, power consumption too high, defective electrical starting.

Key durability issues

Standard of initial specification i.e. correct fan for correct application and location.

Standard of manufacture and quality of materials.

Quality of installation and commissioning .

Incorrect ambient temperature or excessive temperature swings.

Number of start/stops being exceeded for a given period.

Incorrect alignment.

Duty exceeds that which was originally specified.

High rotational speeds – directly affect the noise level, the amount of wear and the life of the fan/system.

Notes

Durability depends not only on the operating environment but also on the use of the fan and its drive motor. Drive motors mounted out of the air stream (or those that are protected in the air stream) have greater reliability. Motors installed away from the air stream reduce turbulence and raise efficiency.

Regreasable bearings should not be over-greased – the presence of a grease relief/release valve will help to prevent over-greasing.

The use of spherical-roller bearings is recommended, as opposed to sleeve bearings.

The Fan Manufacturers' Association were instrumental in setting up the Eurovent/CECOMAF certification scheme and many of the FMA members are participating in it. This scheme ensures that the claimed performance figures of manufacturers have been verified against BS 848: Part1.

Over-rating of motor windings by up to 20% may increase the durability.

Sealed (maintenance free) bearings typically have an operating life of between 15000 and 25000 hours.

Regular cleaning of dust from impellers of in-line fans can extend bearing life. Frequency of cleaning will depend upon amount of dust/impurities in the air.

HVAC FANS – DIRECT DRIVEN *(continued)*

AXIAL FLOW FANS

YEARS	DESCRIPTION	INSPECTION	MAINTENANCE
15	Axial flow fans manufactured and tested to BS 848, with totally enclosed fan cooled three-phase squirrel cage induction motor with rolling element regreasable bearings (to BS ISO 15 and BS 6107). Motor protected to IP55 standard or higher and manufactured to BS 2048, BS 4999, BS 5000 and BS EN 60034. Motor fitted with thermistor winding temperature protection.	All fans to be inspected in accordance with maunfacturer's requirements, to include (typically): **Three monthly:** inspect impeller and motor bearings (unless 'sealed for life' type). **Six monthly:** check flexible conduit connections and wiring, check impeller variable pitch mechanism, check operation of non-return flap. Inspection of fan wheels, sheaves and bearing collars, mounting bolts and guide vanes. Check operation of automatic changeover controls. **Annually:** inspection of impeller and spinnings. Inspection of motor assembly.	Maintenance in accordance with manufacturer's requirements and HVCA specification, to include (typically): **Three monthly:** clean impeller, lubrication of motor bearings (unless 'sealed for life' type). **Six monthly:** clean belt guards, adjustment of drives and belts, fan wheels, sheaves and bearing collars. **Annually:** cleaning of impeller and spinnings.
10	Axial flow fans manufactured and tested to BS 848, with 240V 50Hz single-phase induction motor with output not exceeding 0.37 kW. Motor manufactured to BS 2048, BS 4999 and BS 5000: Part 11. Motor fitted with thermistor winding temperature protection.		
5	Axial flow fans manufactured and tested to BS 848, with motor manufactured to BS 4999 and BS 5000 with protection to IP 22. Motor not fitted with thermistor winding temperature protection.		
U	Unclassified i.e. axial flow fans not manufactured/tested to the above standards.		

PROPELLER FANS

YEARS	DESCRIPTION	INSPECTION	MAINTENANCE
10	Propeller fans manufactured and tested to BS 848, with totally enclosed fan cooled three-phase squirrel cage induction motor. Motor protected to IP55 standard or higher and manufactured to BS 2048, BS 4999, BS 5000 and BS EN 60034. Motor fitted with thermistor winding temperature protection.	All fans to be inspected in accordance with maunfacturer's requirements, to include (typically): **Six monthly:** inspect impeller and motor bearings (unless 'sealed for life' type), check security of mounting fixings, check flexible conduit connections and wiring, check operation of non-return flaps, protection guards and automatic shutters.	Inspection and maintenance in accordance with manufacturer's requirements and HVCA specification, to include (typically): **Six monthly:** clean impeller, lubricate the motor bearings (unless 'sealed for life' type).
5	Propeller fans manufactured and tested to BS 848, with motor manufactured to BS 4999 and BS 5000 with protection to IP 22. Motor not fitted with thermistor winding temperature protection.		
U	Unclassified i.e. propeller fans not manufactured/tested to the above standards.		

Adjustment factors

Dual fan system with automatic or managed changeover: +5 years.

Impellers of insufficient standard for intended use, or not properly balanced: –10 years (or life assessment of 5 years, whichever is greater).

Sealed-for-life bearings used instead of re-greasable bearings: –5 years (20 & 15 year life fans only).

Manufacturing process not quality assured i.e. to ISO 9000 series: –5 years (20, 15 and 10 year life fans only).

The above factors are not cumulative: the factor that is the largest should be applied.

Assumptions – Design & Installation

Fans are to be 'type' tested in accordance with the requirements of BS 848: Parts 1 and 2, and must be selected to deliver the required air volume flow rate. Suitable match to be made between expected usage and related duty and the system type in which the fan is installed.

Fan selection must be appropriate for intended use, i.e. suitability for operation in corrosive environments or in presence of abrasive airborne particles.

Steel casings, frames etc. to be suitably protected against corrosion, e.g. by hot dip galvanizing to BS 729, or aluminum-zinc coating, or appropriate polymeric or stoved enamel coating. Impellers must also be suitably protected. Glass fibre reinforced polymeric impellers are available for use in certain aggressive environments.

Motor enclosures to have a minimum IP (protection) to BS EN 60529 as indicated.

Fan bearings to be suitable for the size, speed, load and discharge angle of the fan.

Check that impeller has adequate clearance and that all local ductwork is properly supported.

Installation in strict accordance with manufacturer's instructions and relevant BSRIA/CIBSE/HVCA/FMA/HEVACA/other guidance/codes.

Foundations and fixing points must suit the dynamic load and frequency of the fan.

Assumptions – Commissioning

Commissioning in strict accordance with manufacturer's instructions and BSRIA Application Guides/CIBSE Commissioning Codes/other appropriate guidance, such as that from the Fan Manufacturers' Association.

Key failure modes

Fans connected to mains supply with wrong electrical characteristics. (Note that some types of fan installed such that they revolve in the wrong direction will still deliver a volume of air, albeit less than the design volume).

Motors not continuously rated.

Corrosion due to inadequate protection/inappropriate selection for intended use/environment.

Fans used/placed in incorrect atmosphere i.e. air handling fans for 'normal' atmospheres used in environment polluted by dust, soot, flour, sawdust etc.

Bearings failure, e.g. due to uneven build up of dust on impellers, causing fan to become out of balance.

Motor failures (N.B. this is a common cause of fan failure).

Impeller failure due to fatigue.

Too much/too little airflow, air surging, excessive noise and vibration, power consumption too high, defective electrical starting.

Key durability issues

Standard of initial specification i.e. correct fan for correct application and location.

Standard of manufacture and quality of materials.

Quality of installation and commissioning .

Incorrect ambient temperature or excessive temperature swings.

Number of start/stops being exceeded for a given period.

Incorrect alignment.

Duty exceeds that which was originally specified.

High rotational speeds – directly affect the noise level, the amount of wear and the life of the fan/system.

Vibration levels should not be greater than 4.5 mm per second.

Other Notes

Durability depends not only on the operating environment but also on the use of the fan and its drive motor. Drive motors mounted out of the air stream (or those that are protected in the air stream) have greater reliability. Motors installed away from the air stream reduce turbulence and raise efficiency.

Regreasable bearings should not be over-greased – the presence of a grease relief/release valve will help to prevent over-greasing.

The use of spherical-roller bearings is recommended, as opposed to sleeve bearings.

The Fan Manufacturers' Association were instrumental in setting up the Eurovent/CECOMAF certification scheme and many of the FMA members are participating in it. This scheme ensures that the claimed performance figures of manufacturers have been verified against BS 848: Part1.

Over-rating of motor windings by up to 20% may increase the durability.

Sealed (maintenance free) bearings typically have an operating life of between 15000 and 25000 hours.

Regular cleaning of dust from impellers of in-line fans can extend bearing life. Frequency of cleaning will depend upon amount of dust/impurities in the air.

HVAC FANS – BELT DRIVEN

CENTRIFUGAL FANS

YEARS	DESCRIPTION	INSPECTION	MAINTENANCE
20	Centrifugal fans manufactured and tested to BS 848.	All fans to be inspected in accordance with maunfacturer's requirements, to include (typically):	Maintenance in accordance with manufacturer's requirements and HVCA specification, to include (typically):
U	Unclassified i.e. centrifugal fans not manufactured/tested to the above standards.	**Six monthly:** inspection of fan/motor bearings (unless 'sealed for life' type), inspection of drives and belts, fan wheels, sheaves and bearing collars, mounting bolts, guide vanes and guards. Check operation of automatic changeover controls. **Annually:** inspection of impeller and spinnings and inspection of motor assembly.	**Six monthly:** lubrication of fan/motor bearings (unless 'sealed for life' type), adjustment of drives and belts, fan wheels, sheaves and bearing collars, mounting bolts, guide vanes and guards. **Annually:** cleaning of impeller and spinnings.

MIXED FLOW FANS

YEARS	DESCRIPTION	INSPECTION	MAINTENANCE
20	Mixed flow fans manufactured and tested to BS 848.	As above.	As above.
U	Unclassified i.e. mixed flow fans not manufactured/tested to the above standards.		

AXIAL FLOW FANS

YEARS	DESCRIPTION	INSPECTION	MAINTENANCE
20	Axial flow fans manufactured and tested to BS 848.	All fans to be inspected in accordance with maunfacturer's requirements, to include (typically):	Maintenance in accordance with manufacturer's requirements and HVCA specification, to include (typically):
U	Unclassified i.e. axial flow fans not manufactured/tested to the above standards.	**Three monthly:** inspect impeller and motor bearings (unless 'sealed for life' type). Inspect the belt drive. **Six monthly:** check flexible conduit connections and wiring, check impeller variable pitch mechanism, check operation of non-return flap. Inspection of drives and belts, fan wheels, sheaves and bearing collars, mounting bolts, guide vanes and guards. Check operation of automatic changeover controls. **Annually:** inspection of impeller and spinnings. Inspection of	**Three monthly:** clean impeller, lubrication of motor bearings (unless 'sealed for life' type). Adjustment of belt drives. **Six monthly:** clean belt guards, adjustment of drives and belts, fan wheels, sheaves and bearing collars, mounting bolts, guide vanes and guards. **Annually:** cleaning of impeller and spinnings.

PROPELLER FANS

YEARS	DESCRIPTION	INSPECTION	MAINTENANCE
10	Propeller fans manufactured and tested to BS 848.	All fans to be inspected in accordance with maunfacturer's requirements, to include (typically):	Inspection and maintenance in accordance with manufacturer's requirements and HVCA specification, to include (typically):
U1	Unclassified i.e. propeller fans not manufactured/tested to the above standards.	**Six monthly:** inspect impeller and motor bearings (unless 'sealed for life' type), check security of mounting fixings, check flexible conduit connections and wiring, check operation of non-return flaps, protection guards and automatic shutters, inspection of drive belts	**Six monthly:** clean impeller, lubricate the motor bearings (unless 'sealed for life' type), adjust any drive belts

Adjustment factors

Dual fan system with automatic or managed changeover: +5 years.

Impellers of insufficient standard for intended use, or not properly balanced: –10 years (or life assessment of 5 years, whichever is greater).

Sealed-for-life bearings used instead of re-greasable bearings: –5 years (20 & 15 year life fans only).

Manufacturing process not quality assured i.e. to ISO 9000 series: –5 years (20, 15 and 10 year life fans only).

The above factors are not cumulative: the factor that is the largest should be applied.

Assumptions – Design & Installation

Fans are to be 'type' tested in accordance with the requirements of BS 848: Parts 1 and 2, and must be selected to deliver the required air volume flow rate.

Fan selection must be appropriate for intended use, i.e. suitability for operation in corrosive environments or in presence of abrasive airborne particles. Suitable match to be made between expected usage and related duty and the system type in which the fan is installed.

Steel casings, frames etc. to be suitably protected against corrosion, e.g. by hot dip galvanizing to BS 729, or aluminum-zinc coating, or appropriate polymeric or stoved enamel coating. Impellers must also be suitably protected Glass fibre reinforced polymeric impellers are available for use in certain aggressive environments.

Belt drives to comply with BS 3790.

Fan bearings to be suitable for the size, speed, load and discharge angle of the fan.

Check that impeller has adequate clearance and that all local ductwork is properly supported.

Installation in strict accordance with manufacturer's instructions and relevant BSRIA/CIBSE/HVCA/FMA/HEVACA/other guidance/codes.

Foundations and fixing points must suit the dynamic load and frequency of the fan.

Assumptions – Commissioning

Commissioning in accordance with manufacturer's instructions and BSRIA Application Guides/CIBSE Commissioning Codes/other appropriate guidance, such as that from the Fan Manufacturers' Association.

Key failure modes

Fans connected to mains supply with wrong electrical characteristics. (Note that some types of fan installed such that they revolve in the wrong direction will still deliver a volume of air, albeit less than the design volume).

Motors not continuously rated.

Corrosion due to inadequate protection/inappropriate selection for intended use/environment.

Fans used/placed in incorrect atmosphere i.e. air handling fans for 'normal' atmospheres used in environment polluted by dust, soot, flour, sawdust etc.

Bearings failure: N.B. this is particularly common in belt driven fans due to the pull exerted by the belt/pulley assembly on a proportion of the bearings.

Belt and sheave failure, e.g. broken belt, misaligned belt, loose belt or pulley.

Out-of-balance impeller. Impeller failure due to fatigue.

Too much/too little airflow, air surging, excessive noise and vibration, power consumption too high, defective electrical starting, belt and sheave failure.

Key durability issues

Standard of initial specification i.e. correct fan for correct application and location.

Standard of manufacture and quality of materials.

Quality of installation and commissioning .

Incorrect ambient temperature or excessive temperature swings.

Number of start/stops being exceeded for a given period.

Incorrect alignment, incorrect belt tension. Incorrect belt and pulley selection.

Duty exceeds that which was originally specified.

High rotational speeds – directly affect the noise level, the amount of wear and the life of the fan/system.

Vibration levels should not be greater than 4.5 mm per second.

Notes

Durability depends not only on the operating environment but also on the use of the fan and its drive motor.

Drive motors mounted out of the air stream (or those that are protected in the air stream) have greater reliability. Motors installed away from the air stream reduce turbulence and raise efficiency.

Regreasable bearings should not be over-greased – the presence of a grease relief/release valve will help to prevent over-greasing.

The use of spherical-roller bearings is recommended, as opposed to sleeve bearings.

The Fan Manufacturers' Association were instrumental in setting up the Eurovent/CECOMAF certification scheme and many of the FMA members are participating in it. This scheme ensures that the claimed performance figures of manufacturers have been verified against BS 848: Part1.

Over-rating of motor windings by up to 20% may increase the durability.

Sealed (maintenance free) bearings typically have an operating life of between 15000 and 25000 hours.

Regular cleaning of dust from impellers of in-line fans can extend bearing life. Frequency of cleaning will depend upon amount of dust/impurities in the air.

DISTRIBUTION DUCTWORK

Scope

This component study provides data on distribution ductwork for ventilation and air conditioning systems in non-domestic building types.

The following component sub-types are included within this section:

Component Sub-type	Page
Rigid ductwork	63
Proprietary fireproof ductwork	65
Flexible ductwork	66

Standards cited

BS 476	Fire tests on building materials and structures.
Part 24:1987	Method of determination of the fire resistance of ventilation ducts.
BS 1449	Steel plate, sheet and strip.
Part 2: 1983	Specification for stainless and heat-resisting steel plate, sheet and strip
BS 5588	Fire precautions in the design and construction of buildings.
Part 9:1999	Code of practice for ventilation and air conditioning ductwork.
BS 5720: 1979	Code of practice for mechanical ventilation and air conditioning in buildings
BS 5925: 1991	Code of practice for ventilation principles and designing for natural ventilation.
BS 8313: 1997	Code of practice for accommodation of building services in ducts.
BS EN 638:1995	Plastic piping and ducting systems.
BS EN 1505:1998	Ventilation for buildings – Sheet metal air ducts and fittings with rectangular cross-section – Dimensions.
BS EN 1506:1998	Ventilation for buildings – Sheet metal air ducts and fittings with circular cross-section – Dimensions.
DD ENV 12097: 1997	Ventilation for buildings. Ductwork. Requirement for ductwork components to facilitate maintenance of ductwork systems.

Other references / information sources

ASHRAE Handbooks:	HVAC applications (1995); HVAC systems and equipment (1996); Fundamentals (1997)
BSRIA Application Guide 89/3	Commissioning of air systems in buildings.
CIBSE TM 8:1983	Design notes for ductwork
CIBSE Guide B:1986	Installation and equipment data – B1 Heating, B2/B3 Ventilation and air conditioning
CIBSE Commissioning Code A: 1996	Air distribution systems
DEO/TSO Spec 037:1998	Air conditioning, air cooling and mechanical ventilation for buildings
Fan Manufacturer's Association Fan and ductwork installation guide	
FETA Eurovent 2/2	Air leakage rate in sheet metal air distribution systems.
FETA Eurovent 2/3	Sheet metal air duct – standard for dimensions.
FETA Eurovent 2/4	Sheet metal air duct – standard for fittings.
HVCA DW 143	A practical guide to ductwork leakage testing
HVCA DW 144	Specification for sheet metal ductwork for low, medium and high pressure/velocity air systems 1998
HVCA DW 151	Plastic ductwork
HVCA DW 191	Glass fibre ductwork
HVCA DW / TM2	Internal cleanliness of new ductwork installations
HVCA Standard Maintenance Specification for Mechanical Services in Buildings – Volumes 1 to 5	
PSA/DEO MEEG 1/06	Air conditioning 1983
PSA/DEO MEEG 1/07	Ventilation 1989
PSA/DEO MEEG 4/06	Air conditioning 1984
PSA/DEO MEEG 4/07	Ventilation 1984

DISTRIBUTION DUCTWORK

RIGID DUCTWORK

YEARS	DESCRIPTION	INSPECTION	MAINTENANCE
35+	Stainless steel to HVCA specification DW 144 and to BS 1449:Part 2.	Periodic inspection for corrosion, dust, dirt, grease, growths etc. Inspection for fire hazards and legionella conditions.	Maintenance in accordance with manufacturer's requirements and HVCA specification, to include (typically): Appropriate maintenance and cleaning programmes (e.g. scheduled wash-down to remove corrosive residues). The frequency of treatment to be appropriate to the local internal and external environmental conditions.
35+	Hot dipped galvanized steel ductwork to HVCA specification DW 144.		
35+	Electroplated galvanized steel ductwork to HVCA specification DW 144.		
35	Mild steel to HVCA specification DW 144.		
35	Aluminium ductwork to HVCA specification DW 144 i.e. alloys 1200, 3103 and 5251 (as specified in BS EN 485, BS EN 515, BS EN 573) and alloy 6082–T6 where high strength is required.	The nature of the natural oxide coating of aluminium provides a protective coating. Inspection of mild steel support sections related to the local internal and external environmental conditions may be required	
20	Glass fibre Resin bonded glass fibre ductwork to HVCA specification DW 191. Generally coated one side with protective film. Board to BS 2972, section 19 and BS 476:Part 7 with a density (excluding any facings) of not less than 48g/m³ or 80g/m³ in preformed circular duct.	If the duct contains mild steel stiffening elements, then periodic inspection of any exposed sections related to the local internal and external environmental conditions may be required. Increased air turbulence can cause break-up of the board and such duct sections may require internal inspection.	
15	PVC-U and polypropylene ductwork to HVCA specification DW 151.	Inspect expansion joints and loops periodically. Frequency of inspection/examination will depend on the condition of the system.	
U	Unclassidied i.e. rigid ductwork not manufactured/installed to the above standards.	As above.	

Adjustment factors

Exposed ductwork located in areas of high traffic and likely to be damaged: –10 years.

Metal ductwork used in areas of high humidity: –10 years (except stainless steel).

Insufficient support brackets or brackets not sufficiently stiff: –10years.

The above factors are not cumulative: the factor that is the largest should be applied.

Assumptions – Design & Installation

Where rigid and flexible duct are connected, compression bands or similar should be the primary method of joint fixing.

All joints to be made using approved methods.

Supports and hanging brackets to be installed to relevant BS/HVCA/other guidance/codes.

Frequency of cleaning to be appropriate to local internal and external environmental conditions.

Installation in strict accordance with manufacturer's instructions and relevant BSRIA/CIBSE/HVCA/ other guidance/codes.

PVC-U and polypropylene have considerable coefficients of expansion. It is essential that adequate provision for movement is made during installation.

Assumptions – Commissioning

Commissioning in accordance with manufacturer's instructions and the appropriate industry guidance e.g. CIBSE Commissioning Code A: 1996 Air distribution systems.

Key failure modes

Leakage, caused by seal failure between sections and also splitting of the joint seams. Commonly caused by poor workmanship and/or incorrect installation

Ducting splits due to vibration caused by air turbulence and transmitted vibration from machines. Usually due to under-specification or change of use

Excessive noise and vibration

Corrosion in damp/humid environments.

Key durability issues

Ductwork is durable and provided it is specified and connected correctly, can be expected to be durable.

Most durability problems are caused by external physical or internal chemical/biological damage to ductwork

A build-up of dust and oil products will result in a fire hazard.

Standard of initial specification, manufacture and quality of materials.

Quality of installation and commissioning, maintenance and inspection regimes and personnel.

DISTRIBUTION DUCTWORK

PROPRIETARY FIREPROOF DUCTWORK

YEARS	DESCRIPTION	INSPECTION	MAINTENANCE
35+	Vermiculite board to BS 476:Part: 24 and ISO 6944.	Periodic inspection for corrosion, dust, dirt, grease, growths etc. Inspection for fire hazards and legionella conditions.	Maintenance in accordance with manufacturer's requirements and HVCA specification, to include (typically): Appropriate maintenance and cleaning programmes (e.g. scheduled wash-down to remove corrosive residues). The frequency of treatment to be appropriate to the local internal and external environmental conditions.
35+	Calcium silicate board to BS 476:Part: 24 and ISO 6944.		
35+	Stainless steel clad cementitious board composites, to BS 476:Part 24 and ISO 6944.		
35	Hot dip galvanized steel clad cementitious board composites, to BS 476:Part 24 and ISO 6944.		
U	Unclassified, ie proprietary fireproof ductwork not manufactured/installed to the above standards.		

Note: this table includes products that are designed to provide fireproof ducts, as opposed to ancillary surface coverings that are used to provide standard ductwork with a certain fire rating.

Adjustment factors

Vermiculite or calcium silicate board unprotected and installed in exposed areas (loading bays, hangers etc): –10 years.

Insufficient support brackets or brackets not sufficiently stiff: –10 years.

Metal faced ductwork used in areas of high humidity: –10 years (except stainless steel).

The above factors are not cumulative: the factor that is the largest should be applied.

Assumptions – Design & Installation

Ducts may be pre-assembled in the factory or assembled on site.

All joints are screwed or stapled, and are sealed with proprietary filler or tape.

In the case of steel clad cementitious board composites, the covering function is to provide support to the board. The steel cover is available in galvanized or stainless finish, depending on the durability requirements.

Vermiculite and calcium silicate board ducting should be installed when the building envelope is weather tight to prevent excessive take-up of airborne moisture.

Assumptions – Commissioning

Commissioning in accordance with manufacturer's instructions and the appropriate industry guidance e.g. CIBSE Commissioning Code A: 1996 Air distribution systems.

Key failure modes

If the duct material receives a severe water drenching, it may cause collapse. Therefore, consideration should be given to areas where fire hoses may be used as primary fire-fighting equipment or where the long-term integrity of the duct is a requirement under fire conditions.

Corrosion of metal facings in damp/humid environments.

Key durability issues

Consideration should be given to movement joints where ducts pass through walls. The inherent rigidity of both components could cause fracture of the duct where movement occurs. Such movement can be restricted by good hanger detailing.

Vermiculite and calcium silicate board have good impact and moisture resistance and are also resistant to low concentrations of acids, alkalis and bleaching agents. They can absorb water, causing a loss of strength.

Steel clad cementitious board has also got very high impact resistance.

DISTRIBUTION DUCTWORK

FLEXIBLE DUCTWORK			
YEARS	DESCRIPTION	INSPECTION	MAINTENANCE
25	Stainless steel strip, corrugated, spirally wound with crimped joints, to BS 476:Parts 6, 7 and 20, and to HVCA specification DW 144.	Periodic inspection for corrosion, dust, dirt, grease, growths etc. Inspection for fire hazards and legionella conditions.	Maintenance in accordance with manufacturer's requirements and HVCA specification, to include (typically): Appropriate maintenance and cleaning programmes (e.g. scheduled wash-down to remove corrosive residues). The frequency of treatment to be appropriate to the local internal and external environmental conditions.
20	Galvanised mild steel strip, corrugated, spirally wound with crimped joints, to BS 476:Parts 6, 7 and 20, and to HVCA specification DW 144.		
20	Aluminium strip, corrugated, spirally wound with crimped joints, to BS 476:Parts 6, 7 and 20, and to HVCA specification DW 144.		
20	Multi-ply aluminium with polyester fabric laminate bonded to high tensile steel wire helix, insulated/ uninsulated, to BS 476:Parts 6, 7 and 20, and to HVCA specification DW 144.		
20	Multi-ply aluminium bonded to high tensile steel wire helix, insulated/uninsulated, to BS 476:Parts 6, 7 and 20, and to HVCA specification DW 144.		
10	PVC coated glass fibre fabric bonded to high tensile steel wire helix, uninsulated, to BS 476: Parts 6, 7 and 20, and to HVCA specification DW 144.		
U	Unclassified, ie flexible ductwork not manufactured/installed to the above standards.		

Adjustment factors

Flexible duct not sufficiently supported or installed too short/long: –5 years.

Assumptions – Design & Installation

Provision of support brackets as required by manufacturer.

Note that flexible ductwork is generally used to connect between main duct runs and end terminals (distribution and extract grilles), and to connect between units where the access presents difficult rigid routing problems. Duct runs should be kept as short as possible and fully extended).

The duct consists of metal foils, spirally wound and corrugated without a support structure or metal foil wrapped or plastics cast over a high tensile steel helix. The helix maintains the unit's shape and the covering provides rigidity and may have properties such as fire resistance, pressure resistance and wear resistance. Sound and heat insulation can be achieved by the addition of an inner sandwich-type layer of fibreglass materials.

To prevent loss of air flow and excessive wear of the duct covering material it is important that the components are matched to the air flow rates expected.

Assumptions – Commissioning

Commissioning in accordance with manufacturer's instructions and the appropriate industry guidance e.g. CIBSE Commissioning Code A: 1996 Air distribution systems.

Key failure modes

Damage due to impact or mishandling during servicing of end unit.

Key durability issues

If correctly specified, installed and handled, very few durability problems are likely to occur.

AIR FILTERS FOR HVAC SYSTEMS

Scope

This section provides data on the most common types of intake air filters for HVAC systems. It includes data on filter housings (i.e. the permanent framework into which the filter fits), reloadable filter frames, and on primary, secondary and high efficiency (HEPA) filter media.

Vee-mat, viscous coated, electrostatic, carbon and grease filters are beyond the scope of this study.

The following component sub-types are included within this section:

Standards cited

BS 729:1971	Specification for hot dip galvanized coatings on iron and steel articles
BS 1474:1987	Specification for wrought aluminium and aluminum alloys for general engineering purposes: bars, extruded round tubes and sections.
BS 3928:1969	Method for sodium flame test for air filters (other than for air supply to I.C. engines and compressors
BS EN 779:1993	Particulate air filters for general ventilation. Requirements, testing, marking.
BS EN 1822	High efficiency air filters (HEPA and ULPA)
Part 1:1998	Classification, performance testing, marking.
Part 2:1998	Aerosol production, measuring equipment, particle counting statistics.
Part 3:1998	Testing flat sheet filter media
BS EN 10142:1991	Specification for continuously hot-dip zinc coated low carbon steel sheet and strip for cold forming: technical delivery conditions.

Other references/information sources

EUROVENT 4/4	Sodium chloride aerosol test for filters using flame photometric technique
EUROVENT 4/8	In situ leak testing of high efficiency filters in clean spaces (DOP aerosol test)
EUROVENT 4/9	Method of testing air filters used in general ventilation for determination of fractional efficiency
EUROVENT 4/10	In situ determination of fractional efficiency of general ventilation filters.
BSRIA	Selection guide SG7/91: Air filters – a selection guide.

AIR FILTERS FOR HVAC SYSTEMS

FILTER HOUSINGS

YEARS	DESCRIPTION	INSPECTION	MAINTENANCE
25	Extruded aluminium alloy sections to BS 1474.	Periodic inspection of edge seals/gaskets. Inspect when changing filters.	Replacement of edge seals/gaskets as necessary.
25	Hot rolled/cold formed mild steel sections, hot dip galvanised to BS 729 after all fabrication, minimum zinc coating weight 450 g/m³		
25	Cold formed mild steel sections, hot dip galvanized before fabrication to BS EN 10142, minimum zinc coating weight 450 g/m³		
20	Hot rolled/cold formed mild steel sections, hot dip galvanised to BS 729 after all fabrication, minimum zinc coating weight 275 g/m³		
20	Cold formed mild steel sections, hot dip galvanized before fabrication to BS EN 10142, minimum zinc coating weight 275 g/m³		
20	Hot rolled/cold formed mild steel sections, electro-phoretic paint finish.		
U	Unclassified, i.e. aluminium or mild steel, not to relevant standards.		

RELOADING FILTER FRAMES

YEARS	DESCRIPTION	INSPECTION	MAINTENANCE
10	Mild steel frames with retaining mesh, hot dip galvanised to BS 729 after all fabrication, minimum zinc coating weight 275 g/m³	Periodic inspection of edge seals/gaskets. Inspect when changing filters.	Replacement of edge seals/gaskets as necessary.
10	Mild steel frames with retainig mesh, hot dip galvanised before fabrication to BS EN 10142, minimum zinc coating weight 275 g/m³		
5	Mild steel frames with retaining mesh, electro-phoretic paint finish.		
U	Unclassified, i.e. reloadable mild steel frames, not to relevant standards.		

Adjustment factors

Filter housings directly exposed to the external environment: −10 years (mild steel)

Filter housings directly exposed to the external environment: −5 years (aluminium)

Assumptions – Design & Installation

Filter housing to be compatible with filter frame, fixings and with surrounding ductwork, i.e. to avoid bimetallic corrosion.

Assumptions – Commissioning

None.

Key failure modes

Corrosion.

Impact damage.

Damage during maintenance / replacement of filter.

Key durability issues

Environmental conditions, particularly dust load but also temperature, humidity and corrosive atmospheres.

Degree of corrosion protection.

Hours of AHU operation, airflow rate and proportion of outdoor air.

PRIMARY FILTERS

RURAL (MTHS)	URBAN (MTHS)	DESCRIPTION	INSPECTION	MAINTENANCE
120	60	Washable filters, class G1 to G4 to BS EN 779. Reticulated polymeric foam filter media.	Filters should be subjected to periodic:	Note that the lives given are for general guidance only. Disposable filters should be replaced, and washable filters cleaned either:
120	60	Washable filters, class G1 to G4 to BS EN 779. Stainless steel mesh filter media.	• visual inspection of filter housing for damage and corrosion	
120	60	Washable filters, class G1 to G4 to BS EN 779. Aluminium mesh filter media.		• at predetermined intervals that reflect the operating environment and/or planned maintenance cycles, or
60	30	Washable filters, class G1 to G4 to BS EN 779. Galvanised steel mesh filter media.	• visual inspection of seals and fasteners for damage	
6	3	Disposable panel filters, class G1 to G4 to BS EN 779. Glass fibre filter media.	• visual inspection of filters for damage	• when the efficiency (as measured periodically by pressure testing) falls below a defined level.
6	3	Disposable panel filters, class G1 to G4 to BS EN 779. Polypropylene/ polyester/ nylon/ aramid/ PVC/ PTFE filter media.	• testing for differential pressure and operation.	Replacement of seals/gaskets as required.
6	3	Disposable filter rolls/pads (for use in reloadable frames), class G1 to G4 to BS EN 779. Glass fibre, polypropylene/ polyester/ nylon/ aramid/ PVC/ PTFE filter media.		
U	U	Unclassified, i.e. washable/disposable filters, not to BS EN 779.		

SECONDARY FILTERS

RURAL (MTHS)	URBAN (MTHS)	DESCRIPTION	INSPECTION	MAINTENANCE
12	6	Multi-pocket/multi-layer bag filter media, class F5 to F9 to BS EN 779, with galvanised/ protective coated steel frame/header.	Filters should be subjected to periodic:	Note that the lives given are for general guidance only. Filters should be replaced either:
12	6	Rigid bag filter media, class F5 to F9 to BS EN 779, with galvanised/protective coated steel frame/header.	• visual inspection of filter housing for damage and corrosion	• at predetermined intervals that reflect the operating environment and/or planned maintenance cycles, or
12	6	Pleated (continuous length) filter, class F5 to F9 to BS EN 779, with galvanised/protective coated steel frame/header.	• visual inspection of seals and fasteners for damage	• when the efficiency (as measured periodically by pressure testing) falls below a defined level.
U	U	Unclassified, i.e. secondary filters, not to BS EN 779.	• visual inspection of filters for damage	Replacement of seals/gaskets as required.
			• testing for differential pressure and operation.	

HIGH EFFICIENCY (HEPA) FILTERS

RURAL (MTHS)	URBAN (MTHS)	DESCRIPTION	INSPECTION	MAINTENANCE
24	12	Pleated (continuous length) HEPA filters to BS EN 1822, BS EN 779, or BS 3928, with galvanised/protective coated steel case.	Filters should be subjected to periodic:	Note that the lives given are for general guidance only. Filters should be replaced either:
24	12	Pleated (continuous length) HEPA filters to BS EN 1822, BS EN 779, or BS 3928, with wooden case.	• visual inspection of filter housing for damage and corrosion	• at predetermined intervals that reflect the operating environment and/or planned maintenance cycles, or
U	U	Unclassified, i.e. HEPA filters, not to BS EN 1822, BS EN 779 or BS 3928.	• visual inspection of seals and fasteners for damage	• when the efficiency (as measured periodically by pressure testing) falls below a defined level.
			• visual inspection of filters for damage	Replacement of seals/gaskets as required.
			• testing for differential pressure and operation.	

Note on the service lives of air filters

The lives quoted are a general guide only and should not be taken as definitive. In practice, the service lives of air filters are dependent upon the specific environment and service conditions, and can vary considerably from location to location. There are also variations in the dust holding capacity and hence service life of different filters due to the design and configuration of the filter media, e.g. method of stitching of bag filters, contours and pleating of filter material. These differences are too numerous and complex to quantify in general terms.

Adjustment factors

The lives are based on 8 hours per day, 5 days per week use. Lives of filters used for longer periods should be reduced proportionately.

Assumptions – Design & Installation

Filter frames/seals and filter media are to be installed in strict accordance with manufacturer's instructions.

Filter frames may be of coated/galvanized steel, cardboard or plastics. It is essential to check suitability for the anticipated environment, particularly in areas of high relative humidity.

The correct filter media must be selected to suit the application i.e. efficiency, surface area and environment (dust load, temperature, humidity, corrosive atmospheres).

Washable filter media should be washed using the detergents recommended by the manufacturer, and handled carefully.

Secondary filters must be adequately protected by upstream primary filters.

HEPA filters must be adequately protected by upstream primary and secondary filters.

Air filters must be adequately protected from direct rain, e.g. by the use of weather louvres.

Air filters should not be installed directly downstream of humidifiers.

Adequate access must be provided for inspection and maintenance of filters.

Assumptions – Commissioning

Commissioning of the HVAC system to be carried out in accordance with manufacturer's instructions and any relevant BSRIA/HVCA/CIBSE commissioning guidance. Particular attention should be paid to the fitting and sealing of the filter media within the frame.

Key failure modes

Air pressure drop due to dust particle build-up.

Distortion due to incorrect fitting

Damage due to inadequate installation/maintenance, e.g. filter fitted the wrong way round, filter media trapped in access doors, inadequate edge sealing.

Corrosion or deterioration due to moisture, gaseous atmospheric pollutants.

Biological contamination.

Wear, breakage, corrosion of washable filter media.

Key durability issues

Standard of manufacture and quality of materials.

Standard of installation. Incorrect installation (e.g. installed back to front, damage to edges) is a common cause of poor performance.

Incorrect specification, e.g. incorrect grade of filter for the application, filter size (surface area) not correctly matched to airflow rate.

Environmental conditions, particularly dust load but also temperature, humidity and corrosive atmospheres.

Hours of AHU operation, airflow rate and proportion of outdoor air.

Level of filtration required.

Quality of upstream filters.

Notes

Past experience has shown that air filters are often changed unnecessarily as part of regular maintenance activities, when other parts of the AHU are being repaired, or when air flow problems occur within the AHU. It should be noted that some filter types do not perform at maximum efficiency until they have been in operation for some time, since they rely on some dust build up for optimum performance.

AIR HEATER BATTERIES FOR HVAC SYSTEMS

Scope

This section provides data on ducted air heating appliances, which are normally termed heater batteries. The batteries may be installed in air ductwork, or may be 'packaged' with other elements as part of a prefabricated air-handling unit. This section provides data on the two main types used in HVAC systems: a) steam and hot water-type, and b) electric-type. Heater batteries that use other media such as hot gases or oil to warm the air are excluded from this study.

The following component sub-types are included within this section:

Component Sub-type	Page
Steam and hot water heater batteries	75
Electric heater batteries	77

Standards cited

BS 729: 1971	Specification for hot dip galvanized coatings on iron and steel articles
BS 1387: 1985	Specification for screwed and socketed steel tubes and tubular and for plain end steel tubes suitable for welding or for screwing to BS 21 pipe threads
BS 1449 Part 1:1991	Steel plate, sheet and strip. Carbon and carbon manganese plate, sheet and strip: Specification for cold tolled narrow strip supplied in a range of conditions for heat treatment and general engineering purposes
BS 2871	Specification for copper and copper alloys. Tubes.
BS 3606: 1992 Part 3: 1989:	Specification for steel tubes for heat exchangers Specification for steel, copper alloy and composite flanges
BS 4504 Part 3:1989	Circular flanges for pipes, valves and fittings (PN designated)
BS 5141 Part 2:1977	Specification for air heating and cooling coils Method of test for rating of heating coils
BS 5720: 1979	Code of Practice for mechanical ventilation and air conditioning in buildings
BS 7671: 1992	Requirements for electrical installations. IEE wiring regulations. 16th edition.
BS 7797: 1998	Specification for industrial duct heaters using metal-sheathed elements
BS 8313: 1997	Code of practice for accommodation of building services in ducts
BS EN 754 Part 7:1998	Aluminium and aluminium alloys. Cold drawn rod/bar and tube. Seamless tubes, tolerances on dimensions and form.
BS EN 1057:1996	Copper and copper alloys. Seamless, round copper tubes for water and gas in

sanitary and heating applications

BS EN ISO 9001:1994	Quality systems. Model for quality assurance in design, development, production, installation and servicing

Other references/information sources

ASHRAE Handbooks:	HVAC applications (1995); HVAC systems and equipment (1996); Fundamentals (1997)
BSRIA	Commissioning of air systems in buildings (1989)
CIBSE Guide B:1986	Installation and equipment data – B1 Heating, B2/B3 Ventilation and air conditioning (1986)
CIBSE	Commissioning Code A – Air distribution systems (1996)
DEO Specification 037	Air conditioning, air cooling and mechanical ventilation for buildings (1998)
Electricity at Work Regulations 1989	
IEE	Code of practice for in-service inspection and testing of electrical equipment
Fan Manufacturer's Association Fan and ductwork installation guide (1993)	
HVCA	Standard Maintenance Specification for Mechanical Services in Buildings Pressure Systems and Transportable Gas Regulations (1989)
PSA/DEO MEEG 1/06	Air conditioning (1983)
PSA/DEO MEEG 4/06	Air conditioning (1984)

AIR HEATER BATTERIES FOR HVAC SYSTEMS

STEAM AND HOT WATER HEATER BATTERIES

YEARS	DESCRIPTION	INSPECTION	MAINTENANCE
25	Heater batteries formed from austenitic stainless steel tube to BS 3606 with fins of copper minimum 0.3mm thick, or aluminium minimum 0.4mm thick. Battery tested to 2.1 Mpa or 1.5 times working pressure. Manufacturer ISO 9001 approved.	Steam and HTHW systems operating at 0.5 bar above atmospheric pressure require mandatory regular inspections, formally undertaken by 'competent' persons, and recorded in accordance with the Pressure Systems and Transportable Gas Regulations 1989. All systems to be subject to annual visual inspection of headers, tube joints and coil faces to locate any leaks, corrosion, fouling, erosion etc.	*Annually:* Remove the build-up of atmospheric dust on the fin surface by vacuum or air-blast techniques. In the event of tube leak or significant failure, the complete coil should be removed for maintenance purposes. Water coils should be adequately drained and blown out for winter operation, to prevent freezing. During winter operation, if the unit is shut down for any period, the steam coils should also be drained.
25	Heater batteries of minimum 0.9mm copper tube (0.7mm LTHW) to BS 2871: Part 3 with fins of copper minimum 0.3mm thick, or aluminium minimum 0.4mm thick. Battery tested to 2.1 Mpa or 1.5 times working pressure. Manufacturer ISO 9001 approved.		
25	Heater batteries formed from ferritic stainless steel tube to BS 3606 with fins of copper minimum 0.3mm thick, or aluminium minimum 0.4mm thick. Battery tested to 2.1 Mpa or 1.5 times working pressure. Manufacturer ISO 9001 approved.		
20	Heater batteries of aluminium tube (Grade 1100) to BS EN 754–7 with fins of copper minimum 0.3mm thick, or aluminium minimum 0.4mm thick. Battery tested to 2.1 Mpa or 1.5 times working pressure. Manufacturer ISO 9001 approved.		
15	Heater batteries of carbon steel tube to BS 3606 with fins of copper minimum 0.3mm thick, or aluminium minimum 0.4mm thick. Battery tested to 2.1 Mpa or 1.5 times working pressure. Steel tube hot dip galvanised to BS 729 (or equivalent corrosion protection). Manufacturer ISO 9001 approved.		
U	Unclassified i.e. heater batteries not formed or tested to the above standards.		

Adjustment factors

The above lives are based on use in low temperature hot water systems. Use with steam or high/medium temperature hot water systems: −5 years.

Installed in adverse (but not severely corrosive) environments: −5 years

The above lives assume 5 day a week 12 hour per day operation. Longer operating durations will reduce the life of the component. Continuous operation (24 hours / 7 day week): −5 years

Note: The above factors are not cumulative: the factor that is the largest should be applied.

Assumptions – Design & Installation

Design and installation of the ventilation system to be in strict accordance with manufacturer's instructions and industry good practice (eg CIBSE/BSRIA/HVCA guidance and relevant MoD standards) and following guidance in BS 8313 and BS 5720, with particular regard to the selection and compatibility of materials in the system. Heating coils to be rated to BS 5141:Part 2.

Copper tubes to be of solid drawn copper, expanded into collars formed on the copper or aluminium fin.

Aluminium fins are not suitable for corrosive atmospheres unless protected, e.g. with a polyester coating.

Adequate provision to be included for the thermal expansion of elements of the heater battery (eg expansion loops, expansion/swing joints, flexible connectors).

Adequate measures (e.g. installing low temperature interlocks, frost thermostats or by draining coils when necessary) should be taken to prevent freezing in the coils.

Heater battery casings are to be provided with appropriate protective finishes (e.g. minimum 275g/m² galvanising) to delay corrosion or damage caused by condensation or moisture carried through the air, where the heaters are used in conjunction with air washers/humidifiers and placed 'downstream' of them. For enhanced corrosion protection, stainless steel, copper, brass or aluminium may be used. Proprietary anti-corrosion coatings are also available.

Adequate water treatment (e.g. softening) must be provided to suit the local water quality, especially when higher temperatures are used.

A strainer/filter must be fitted in the supply line to prevent problems with scale build up.

The heater battery must be selected to suit the particular application, i.e. pressure, temperature, heating media, water quality and air quality (with suitable protective finish, as recommended by manufacturers).

Assumptions – Commissioning

Pre-commissioning cleaning of water systems to be carried out in accordance with industry good practice (e.g. BSRIA AG2/89 & AG8/91).

Commissioning to be in strict accordance with manufacturer's instructions and relevant BSRIA/CIBSE/HVCA/MoD/other guidance/codes.

Key failure modes

Corrosion of heat exchanger coils/tubes/fins. NB: this is by far the most common cause of failure.

Corrosion of casing.

Internal fouling, scaling, corrosion (all water-related) and vibration, leading to tube leakage and cracking.

Erosion of fins due to excessively high water velocity and/or inadequate water treatment.

Failure at the heads of brazed joints, due to water hammer and condense logging.

Airlocked coil.

Reduced flow rate due to partially or fully blocked coil (e.g. due to build-up of corrosion products).

Thermal shock in steam coils, due to rapid cooling caused by condensation of stem.

Key durability issues

Standard of manufacture and quality of materials.

Corrosion resistance and protection of base materials.

Overall water quality and type of water treatment used.

Suitability of protective coatings, i.e. to prevent corrosion in damp/humid environments.

Nature of the heating medium (e.g. hot water, steam, gas or electric heating elements).

Adequate regulation of steam pressure and temperature is essential to prevent failure due to water hammer and thermal shock.

Number of hours of operation.

Notes

There are no standard third party test requirements for heater batteries, but some manufacturers have the coil-type tested by a third party test house, tagging the units with lead seals after the test.

Pressure/leak testing is usually carried out by the manufacturer, but the percentage of total units tested differs greatly.

CIBSE Guide B recommends that water-based batteries are tested with water at 2.1 Mpa or 1.5 times the working pressure, whichever is greatest.

Draining of a hot water system can introduce additional hardness salts and oxygen to the system, thus increasing scaling and corrosion. Isolating valves should be used to keep the section to be drained to a minimum.

Aluminium finned heater batteries are available with a polyester coating for additional corrosion protection in aggressive environments.

AIR HEATER BATTERIES FOR HVAC SYSTEMS *(continued)*

ELECTRIC HEATER BATTERIES

YEARS	DESCRIPTION	INSPECTION	MAINTENANCE
10	Electric heater batteries manufactured and tested in accordance with BS 7797 and installed to BS 7671. Manufacturer ISO 9001 approved.	Electric heater batteries are subject to the Electricity at Work Regulations 1989 and its requirements relating to safe operation and maintenance.	*Annually:* Check scale deposits and corrosion.
U	Unclassified i.e. Electric heater batteries not manufactured and tested in accordance with BS 7797 or not installed to BS 7671.	Inspection and testing requirements stated in IEE 'Code of practice for in-service inspection and testing of electrical equipment' should be followed.	Check the functioning of safety devices. Clean the air side.
		Heater batteries should also be subjected to periodic visual inspection during commissioning (and then annually) to detect any deterioration or damage.	

Adjustment factors

Manufacturing process not quality assured (i.e. to BS EN ISO 9000 series): −5 years

The lives assigned above assume a 5 day per week, 12 hour per day operation. Longer operation will reduce the life of the component. Continuous operation (24 hours / 7 day week): −5 years.

Note: The above factors are not cumulative: the factor that is the largest should be applied.

Assumptions – Design & Installation

Design and installation of the ventilation system to be in strict accordance with manufacturer's instructions and industry good practice (eg CIBSE/BSRIA/HVCA guidance and relevant MoD standards) and following guidance in BS 8313 and BS 5720, with particular regard to the selection and compatibility of materials in the system.

Adequate provision to be included for the thermal expansion of elements of the heater battery (eg expansion loops, expansion/swing joints, flexible connectors).

A high-limit thermostat should be incorporated, which will trip the heater if the temperature rises above a certain limit. This will prevent the battery from being damaged due to overheating.

Heater battery casings are to be provided with appropriate protective finishes (eg minimum 275g/m² galvanizing) to delay corrosion or damage caused by condensation or moisture carried through the air, where the heaters are used in conjunction with air washers / humidifiers and placed 'downstream' of them. For enhanced corrosion protection, stainless steel, copper, brass or aluminium may be used. Proprietary anti-corrosion coatings are also available.

The heater battery must be selected to suit the particular application, with a suitable protective finish, as recommended by the manufacturer.

Assumptions – Commissioning

Commissioning to be in strict accordance with manufacturer's instructions and relevant BSRIA/CIBSE/HVCA/MoD/IEE/other guidance/codes.

Key failure modes

Failure of heating element.

Main contactor de-energised.

Corrosion due to adverse surrounding environment.

Careless storage and/or handling during installation.

Key durability issues

Standard of manufacture and quality of materials.

Quality of handling, installation and commissioning.

Corrosion resistance of base materials.

Hours of operation.

Notes

Life expectancy may be reduced if the electric heating element repeatedly expands and contracts in use. This may be unavoidable due to the application or due to poor control design or due to insufficient staging of large output units

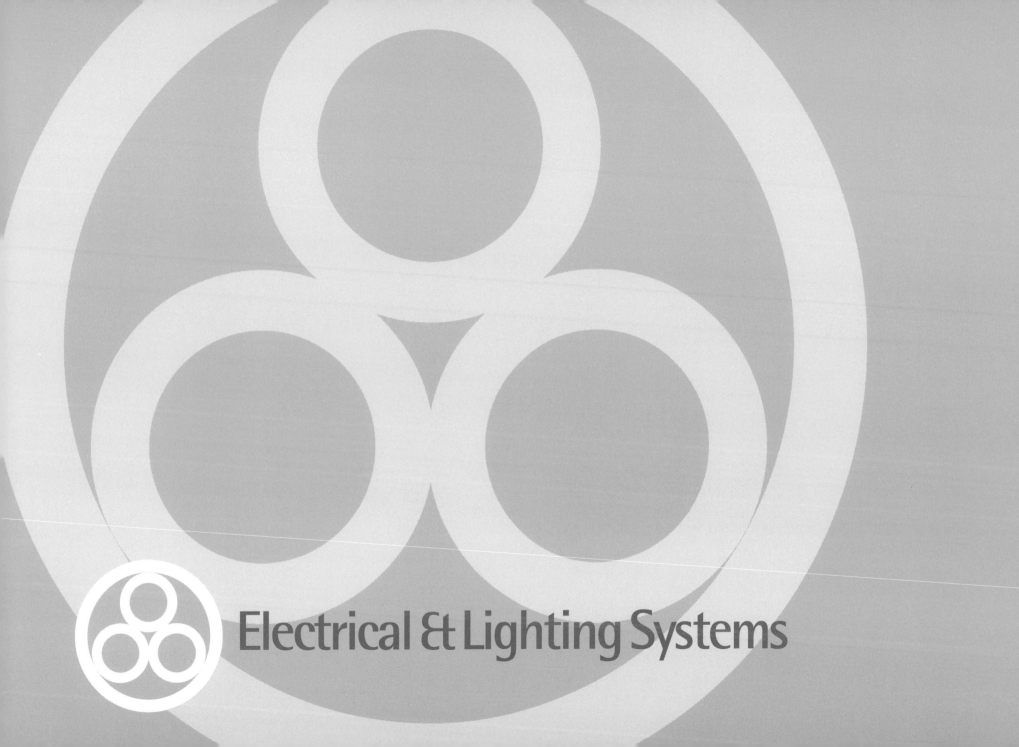

Electrical & Lighting Systems

LOW VOLTAGE CABLES

Scope

This section provides data on low voltage cables, i.e. less than 1kV, for use in non-domestic building types. Medium and high voltage cables, and special cables used for microelectronics, data, communications, fibre optics etc. are excluded from this study.

The following component sub-types are included within this section:

Standards cited

BS 4109:1970	Copper for electrical purposes. Wire for general electrical purposes and for insulated cables & flexible cords.
BS 5467:1997	Cables with thermosetting insulation for electricity supply for rated voltages of up to & inc.600/1000V and up to and including 1900/3300V.
BS 6004:1995	PVC insulated cables (non-armoured) for electric power and lighting.
BS 6207:1995	Mineral insulated cables with rated voltages not exceeding 750V, 2 parts.
BS 6346:1997	600/1000V & 1900/3300 armoured electric cables having PVC insulation.
BS 6387:1994	Specification for performance requirements for cables required to maintain circuit integrity under fire conditions.
BS 6480:1988	Impregnated paper insulated lead or lead alloy sheathed.
BS 6724:1997	Armoured cables having thermosetting insulation with low emission of smoke and corrosive gases when affected by fire.
BS 7211:1998	Thermosetting ins. Cables. (non-armoured) for electric power and lighting. With low emission of smoke/ corrosive gases when affected by fire.
BS 7629:1997	300/500V fire resistant cables with low emission of smoke and corrosive gases when affected by fire – 2 parts.
BS 7671:1992	Requirement for electrical installations. IEE wiring regs. 16th edition.
BS 7846:1996	600/1000V armoured fire-resistant electric cables having low emission of smoke & corrosive gasses when affected by fire.
BS 7889:1997	600/1000V single core unarmoured electric cables having thermosetting insulation.

Other references/information sources

British Approvals Service for Electrical Cables (BASEC) testing/approval

BSRIA: 1997	Safe thermal imaging of electrical systems (up to and including 1000V ac)
BSRIA: 1999	Guidance and the standard specification for thermal imaging of LV electrical installations 1000V ac)
BSRIA TN 19/95: 1995	Rules of thumb
BSRIA TN 8/92: 1992	Small power loads
CIBSE Guide B 10	Electrical power
DEO(W)/DEO TB 59/94:1994	Health and safety – supply voltage – harmonisation throughout the EU – its impact on the MoD estate
DEO(W)/DEO TB 96/18:1996	MoD electricity safety rules and procedures
DEO(W)/TSD SRP 01:1996	Electricity
DEO(W)/DEO TB 97/08:1997	Issue of MoD electricity safety rules and procedures SRP 01
HVCA Standard Maintenance	Specification for Mechanical Services in Buildings – Volumes 5: Electrics in buildings
IEE: BS 7671:1992	Requirements for electrical installation – IEE wiring regulations; 16th edition
PSA/DEO SSME 1	Electrical installations 1985
PSA/DEO MEEG 3/22	Transmission and distribution 1983
PSA/DEO MEEG 3/23	LV electrical installations and equipment 1986
PSA/DEO MEEG 3/24	LV Installations in hazardous areas 1983
PSA/DEO MEEG 6/22	Transmission and distribution 1986
PSA/DEO MEEG 6/23	LV electrical installations and equipment 1984

The Electricity at Work Regulations

LOW VOLTAGE CABLES

MINERAL INSULATED CABLES

YEARS	DESCRIPTION	INSPECTION	MAINTENANCE
35+	Mineral insulated single and multi-cored cables to BS 6207.	Inspect for: physical damage (from people, machinery/furniture/animals); exposure to heat, radiation, oil, corrosion, extremes of Ph, moisture etc; general cleanliness. Check terminations for tightness and deterioration, due to the heating effect of arcing. Periodic testing to include: continuity of live conductors and neutral; earth continuity, insulation resistance between live conductors and to earth; earth loop impedance tests.	The Electricity at Work Regulations assume that a system is safe for life and will only require maintenance attention to sustain integrity. Therefore, maintenance will only be required if the lack of it will result in a danger arising. It follows that periodic inspection and testing of electrical installations is a requirement
35+	Mineral insulated single and multi-cored and plastic sheathed cables to BS 6207.		
U1	Unclassified, i.e. mineral insulated cables, not to BS 6207.		

PAPER INSULATED CABLES

YEARS	DESCRIPTION	INSPECTION	MAINTENANCE
35+	Paper insulated and metal sheathed cables (PILCSWA) to BS 6480.	As above.	As above.
U	Unclassified, i.e. paper insulated cables, not to BS 6480.		

THERMOSETTING PLASTIC INSULATED CABLES

YEARS	DESCRIPTION	INSPECTION	MAINTENANCE
35	Thermosetting plastic insulated single and multi-cored fire performance armoured cable with copper conductors to BS 6387, BS 6724, BS 7629 and BS 7846.	As above.	As above.
35	Thermosetting plastic insulated single and multi-cored fire performance armoured cable with aluminium conductors to BS 6387, BS 6724, BS 7629 and BS 7846.		
35	Thermosetting plastic insulated single and multi-cored armoured cable with copper conductors to BS 5467.		
35	Thermosetting plastic insulated single and multi-cored armoured cable with aluminium conductors to BS 5467.		
30	Thermosetting plastic insulated single and multi-cored non-armoured cable with copper conductors to BS 7211 and BS 7889.		
30	Thermosetting plastic insulated single and multi-cored non-armoured cable with aluminium conductors to BS 7211 and BS 7889.		
U	Unclassified, i.e. thermosetting plastic insulated cables, not to relevant standards.		

THERMOPLASTIC INSULATED CABLES

YEARS	DESCRIPTION	INSPECTION	MAINTENANCE
30	Thermoplastic insulated single and multi-cored armoured cables with copper or aluminium conductors to BS 6346.	As above.	As above.
25	Thermoplastic insulated single and multi-cored armoured cable with copper conductors to BS 6004.		
U	Unclassified, i.e. thermoplastic insulated cables, not to relevant standards.		

Adjustment factors

Manufacturing process not quality assured i.e. to ISO 9000 series: –5 years.

Regular testing and inspection of non-armoured cables: +5 years.

Assumptions – Design & Installation

Cables to be tested and approved by British Approvals Service for Electrical Cables (BASEC).

Cable selection in accordance with BS 7540 (cables not exceeding 450/750V).

Design requirements for systems to be upheld at installation.

Performance and design limitations for cables to be observed at installation and throughout service life.

Thermoplastic insulated cables to be installed where ambient temperature does not exceed 30°C and conductor operating temperature does not exceed 70°C.

Thermosetting insulated cables to be installed where ambient temperature does not exceed 30 ∞C and conductor operating temperature does not exceed 90°C.

Manufacturers' data and BS 7671 should also be consulted.

Armouring to be specified where necessary to provide protection against mechanical damage.

All supports, conduit, trunking etc. to be installed to relevant BS/IEE/HVCA/other guidance/codes.

Installation in strict accordance with manufacturer's instructions and relevant BS/IEE/HVCA/other guidance/codes.

Assumptions – Commissioning

Commissioning of electrical systems in accordance with manufacturer's instructions and the appropriate industry guidance.

Key failure modes

Thermal ageing and loss of electrical insulation resistance.

Overloading of cables, leading to overheating.

Mechanical damage or stress

Chemical attack, e.g. exposure to solvents.

Electrical arcing.

Heat-shrink joint failure.

Failure due to moisture/water.

Key durability issues

Unarmoured PVC cables are not suitable for external use

Cables to be protected from vermin, solvents and aggressive chemicals, grouping / overcrowding and mechanical damage

Operating temperature – high and low temperatures can be detrimental.

Selection of sheathing/armouring appropriate to the location/use.

Notes

Insulating material is selected for its electrical characteristics, whereas sheathing is selected to protect the cable within its expected environment. It will be necessary to consider ambient or operating temperatures, potential for mechanical damage or abrasion, contact with other materials, flexibility, fire, danger from fumes, and contact with liquids e.g. oil, water, marine salt or chemicals.

The most frequently stated reason for cable replacement is due to the obsolescence of the plant and the buildings which they serve. By allowing a proportion of spare current carrying capacity at the design stage, the useful life of cables can be extended.

LAMPS & LUMINAIRES

Scope

This section provides data on common types of lamps and luminaires manufactured for use in commercial and industrial premises. Indoor display and architectural lighting, outdoor lighting, and lighting for specialised uses/locations is beyond the scope of this study.

The following component sub-types are included within this section:

Standards cited

BS 4533	Luminaires. Parts are related to or are identical to BS EN 60598 and IEC 598.
BS 5266	Emergency lighting
Part 1: 1999	Code of practice for the emergency lighting of premises other than cinemas.
BS 7430: 1998	Code of practice for earthing.
BS 7501: 1989	General criteria for the assessment of testing laboratories.
BS 7671: 1992	IEE Wiring Regulations 16th Edition. Requirements for electrical installations.
BS 8206	Lighting for buildings
Part 1: 1985	Code of practice for artificial lighting.
BS EN 45001/BS 7501	General criteria for operation of testing laboratories.
BS EN 55015:1996	Limits and methods of measurement of radio disturbance characteristics of electric lighting and similar equipment.
BS EN 60081:1998	Double capped fluorescent lamps – performance specifications.
BS EN 60192:1993	Specification for low-pressure sodium lamps.
BS EN 60400:1996	Lampholders for tubular fluorescent lamps and starterholders.
BS EN 60598	Luminaires
Part 1:1997	General requirements and tests.
Part 2: (various)	Particular requirements.
Section 1	Specification for fixed general-purpose luminaires.
Section 2.2	Recessed luminaires.
Section 2.6	Luminaires with built-in transformers for filament lamps.
Section 2.19	Specification for air-handling luminaires (safety requirements).
Section 2.22	Specification for luminaires for emergency lighting.
BS EN 60662:1993	Specification for high-pressure sodium vapour lamps.
BS EN 61000 series	Electromagnetic compatibility (EMC).
BS EN 61167:1995	Specification for metal halide lamps.
BS EN 61199:1994	Single-capped fluorescent lamps – safety specifications.
IEC 60188	High-pressure mercury vapour lamps.

Other references / information sources

Association of Short Circuit Testing Authorities (ASTA) Guide to product approval.

Federation of British Electrotechnical and Allied Manufacturers, Associations (BEAMA) / Lighting Industry Federation (LIF): Industry standard for the construction and performance of battery-operated emergency lighting equipment.

CIBSE	Code for interior lighting, 1994.
CIBSE Guidance Note GN4	Lighting requirements of Building Regulations Part L.
CIBSE Lighting Guide LG 1	The industrial environment.
CIBSE Lighting Guide LG 3	The visual environment for display screen use.
CIBSE Lighting Guide LG 7	Lighting for offices.
CIBSE / ECA / HVCA	Standard maintenance specification for services in buildings. Volume 5 Electrics in buildings.
Defence Estates Organisation	Technical Publications Index: June 1997.

MoD Joint Service Publication 315.

MoD Energy Consumption Guide 75:	Energy in MoD establishments.

Electrical Equipment Safety (Safety) Regulations 1994 – The Low Voltage Directive.

Electricity Association guide G 5/4	Limits for harmonics in the UK electricity supply system.
ILIGHT Technical Report 21	Interior high intensity discharge lighting.
LIF Technical Statement No. 10	The handling and disposal of lamps.
PSA/DEO MEEG 3/25, 6/25	Lighting.

LAMPS (COMMERCIAL)

LINEAR FLUORESCENT LAMPS

HRS	DESCRIPTION	INSPECTION	MAINTENANCE
6000–12000	Tri-phosphor lamps to BS EN 60081.	Periodic inspection for signs of deterioration.	Periodic cleaning in order to maintain lighting output.
6000–12000	Multi-phosphor lamps to BS EN 60081.		
6000–12000	Halophosphate lamps to BS EN 60081 (26mm diameter).		
5000–10000	Halophosphate lamps to BS EN 60081 (38mm diameter).		

COMPACT FLUORESCENT (CLF) LAMPS

HRS	DESCRIPTION	INSPECTION	MAINTENANCE
8000–10000	Compact fluorescent lamps to BS EN 61199.	Periodic inspection for signs of deterioration.	Periodic cleaning in order to maintain lighting output.

HIGH INTENSITY DISCHARGE (HID) LAMPS

HRS	DESCRIPTION	INSPECTION	MAINTENANCE
14000–28000	High pressure sodium lamps to BS EN 60662.	Periodic inspection for signs of deterioration.	Periodic cleaning in order to maintain lighting output.
12000–24000	Low pressure sodium lamps to BS EN 60192.		
6000–13000	Metal halide lamps to BS EN 61167.		
14000–28000	High pressure mercury lamps (fluorescent) to IEC 60188.		
6000–12000	High pressure mercury lamps (blended) to IEC 60188.		

Note: The above figures are averages taken from manufacturers' technical information and from Table 3.4 of the CIBSE Code for Interior Lighting, 1994. Reference should be made to the manufacturer's test data.

Adjustment factors

Poor regulation of mains voltage (fluorescent) –25%

Assumed switching frequency of x8 per day increased to x 15 (fluorescent) –25%

Assumed switching frequency of x8 per day increased to x 15 (compact fluorescent only) –40%

Exposure to excessive vibration (sodium only) –25%

Assumptions – Design and manufacture

Published lamp life data to be based on BS EN 60081 test procedures.

Manufacturer should employ test procedures to demonstrate compliance with standards, either in their own laboratory or that of a third party assessor, to BS EN 45001 / BS 7501.

Manufacturer to be a member of an accredited UK trade association or regulating body (such as the Lighting Industry Federation).

Requirements for electrical equipment earthing and sizing of earthing conductors are contained in BS 7430.

Anti-cycling ignitor should be fitted to luminaires for use with high-pressure sodium lamps to prevent nuisance flicker at end of lamp life.

High-pressure sodium lamps with external ignitor have a longer life (generally more than twice that of lamps with internal ignitor).

Lamp types in table are only a sample of those available for the luminaires in this study.

Fluorescent switching per day is set at 8 (once per 3 hours) in BS EN 60081.

Fluorescent optimum light output is at 25°C. A rise or fall of 10°C will reduce light output by 10% and 20°C a reduction of 20% (approximate figures for illustrative purposes).

For fast switching of compact fluorescent lamps, an electronic starter is required. Life of integral starter switches limited to 20,000 operations.

Ensure that lamps fitted with an integral ignitor are not fitted to circuits designed with external ignitors.

Life of high intensity lamps depends on correct matching of lamp with control gear. Original equipment parameters / characteristics to be observed throughout the maintenance life of the luminaire.

Assumptions – Installation and Commissioning

Lamp life is highly dependent upon handling. Manufacturer's instructions for unpacking, installing and exchanging lamps to be strictly observed.

Some manufacturing faults are apparent soon after installation. Post installation site checks are valuable.

Key failure modes

Loss of light output due to ageing.

Loss of gas or vacuum to lamp.

Impact damage.

High resistance or corrosion at lamp connections.

Damage due to poor environmental control i.e. exposure to high temperatures.

Wide variations in mains voltage can significantly affect the lives of fluorescent lamps, as over-voltage causes more rapid evaporation of emitter coating on tungsten filaments and also blackening of inside surface of tube at ends.

Frequent switching (in excess of 15 times per day) will reduce the life of the lamp and the starting circuit, particularly with compact fluorescents.

Key durability issues

Survival rates for high intensity discharge lamps are based on 10 running hours per start. Shorter intervals between starts will reduce lamp life.

Optimum lamp replacement cycles are dependent upon both the decline in lighting output and the probability of lamp failure. The relative weight given to these two factors will depend on the lamp type and on the maintenance policy of the organisation. The CIBSE Code for interior lighting states that mains and low voltage tungsten filament and tungsten halogen lamps generally fail before any significant drop in lighting output. Other lighting types tend to show a significant drop in performance before a significant proportion has failed. Table 3.4 of the CIBSE guide provides generic data on average lamp life and time to 30% drop in output. In most situations, group replacement of lamps at pre-defined intervals based on manufacturer's data will be the most cost effective option.

Light output of phosphor coated mercury discharge lamps falls quickly. Because of the reduction in output, even lamps with a life in excess of 20,000 hours may require replacement after 10,000 hours (depending on local requirements).

Notes

Requirements for protective devices for lighting installations in general are contained in BS 7671 and summarised in BS 7430.

The lamps described in the above table are only a representative sample of some of the most frequently used lamps.

Lamp efficacy and light output vary with supply voltage.

LUMINAIRES

COMMERCIAL LUMINAIRES			
YEARS	DESCRIPTION	INSPECTION	MAINTENANCE
20	Recessed fluorescent luminaire suitable for mounting in suspended ceilings with air handling capability to BS EN 60598–2–19. Stove enamel body, 1,2,3 or 4 lamp ways and with aluminium low maintenance louvre or UV stabilised prismatic or opal diffuser to BS 4533 & BS EN 60598–2–2 with linear or compact fluorescent. Luminaire to meet CIBSE LG 3 Categories 1, 2 or 3.	*Annually:* check installation for damage to luminaires and general deterioration of room illuminance. If necessary check lighting levels with luminance meter against design illuminance. See lamp schedule for details of optimum replacement intervals.	*Annually:* clean luminaire body, reflecting surfaces and refractor with cleaning materials as described by manufacturer. See CIBSE Code for interior lighting for detailed guidance on cleaning methods for different materials.
20	Surface mounted or suspended fluorescent luminaire, lengths 1300 mm to 2000 mm approx, lamps 1200 to 1800 nominal. Stove enamel sheet steel or aluminium body, for one, two, three, or four lamps, with ABS end caps, aluminium louvres and with open top ventilated body, to BS 4533 & BS EN 60598 and to CIBSE LG 3 Categories 1, 2 or 3	Inspect lampholders for deterioration. Inspect lamps for uniformity of type and colour.	*Ten yearly:* replace capacitor, ignitor, and ballast. Replace lampholders if required. Check electrical connections.
20	Batten fluorescent luminaire, lengths 600 mm to 2400 mm, with stove enamel steel or aluminium spine, optional UV stabilised diffuser, and optional louvre of anodised aluminium to BS 4533 & BS EN 60598.		
15	Economy range batten fluorescent luminaire, lengths 600 mm to 2400 mm, with or without diffuser and rolled steel spine to BS 4533 & BS EN 60598–1 & BS EN 60598 – 2–22.		
U	Unclassified i.e. luminaires of unknown material and / or not manufactured to the above standards.		

INDUSTRIAL LUMINAIRES			
YEARS	DESCRIPTION	INSPECTION	MAINTENANCE
20	High bay luminaire to BS 4533 / BS EN 60598–1. Body of die cast aluminium, with glass or aluminium reflector, lamp vertical cap up, open below reflector.	*Annually:* check installation for damage to luminaires and general deterioration of room illuminance. Check lighting levels with luminance meter against design illuminance if necessary. See lamp schedule for details of optimum replacement intervals.	*Annually:* clean luminaire body, reflecting surfaces and refractor with cleaning materials as described by manufacturer. See CIBSE Code for interior lighting for detailed guidance on cleaning methods for different materials.
15	Shallow discharge luminaire for low bay mounting, to BS 4533 / BS EN 60598–1. Stove enamel coated mild steel body and aluminium reflector with lamp horizontal.	Inspect lampholders for deterioration.	Maintenance cleaning cycle can be extended to five yearly, see notes.
U	Unclassified i.e. luminaires of unknown material and/or not manufactured to the above standards.		*Ten yearly*: replace capacitor, ignitor, and ballast. Replace lampholders if required. heck electrical connections.

Adjustment factors

Regular cleaning: +5 years.

Assumptions – Design & manufacture

Manufacturer should employ test procedures to demonstrate compliance with standards, either in their own laboratory or that of a third party assessor to BS EN 45001 / BS 7501.

Products and components to carry the BSI Kitemark or European equivalent for measurement and testing to standard.

Products bearing the ENEC mark indicate compliance with all relevant European standards.

Products must bear the CE mark to indicate compliance with Low Voltage Directive (LVD) and Electromagnetic Compatibility (EMC).

Manufacturer to be a member of an accredited UK trade association or regulating body (such as the Lighting Industry Federation).

Requirements for electrical equipment earthing and sizing of earthing conductors are contained in BS 7430.

Requirements for protective devices for lighting installations in general are contained in BS 7671 and summarised in BS 7430.

Luminaire radio interference limited by BS EN 55015.

Luminaires for emergency lighting use are to comply with BS 5266 and BS EN 60598 section 2–22.

Plastic lampholders to be UV stabilised.

Ballast tap settings to be set to within 3% of local mains supply voltage for high pressure sodium luminaires. If tap settings are inaccurate by ±15% or greater, lamp life can be seriously affected.

Assumptions – Installation & commissioning.

Planned luminaire switching arrangement to be checked post installation.

Design levels of illumination to be checked and recorded prior to hand-over.

STANDBY POWER GENERATING SETS

Scope

This section provides data on commonly used standby power generating sets. These sets are often required in buildings where essential functions (i.e. computer suites, communication centres, security systems, smoke extraction etc) need to be maintained should the mains supply fail. Gas and steam turbine generators, petrol driven generators, generators for combined heat and power, and three phase high voltage generators are beyond the scope of this section.

The following component sub-types are included within this section:

Component Sub-type	Page
Diesel engine prime mover	93
Alternator	95
Starter batteries	97

Standards cited

BS 4999	General requirements for rotating electrical machines.
Part 105:1988	Classification of degrees of protection provided by enclosures for rotating machinery.
Part 140:1987	Specification for voltage regulation and parallel operation of a.c. synchronous generators.
BS 5000	Specification for rotating electrical machines of particular types or for particular applications.
Part 3:1980	Generators to be driven by reciprocating internal combustion engines.
BS 5514	Reciprocating internal combustion engines. Performance.
Part 1:1996	Standard reference conditions, declarations of power, fuel and lubricating oil consumptions and test methods.
Part 4:1997	Speed governing.
Part 5:1979	Torsional vibrations.
BS 6290:	Lead-acid stationary cells and batteries.
Part 2:1984	Specification for lead-acid high performance Plante positive type.
Part 3:1999	Specification for lead-acid pasted positive plate type.
Part 4:1997	Specification for classifying valve regulated types.
BS 7698:	Reciprocating internal combustion engine driven alternating current generating sets.
Part 1:1993	Specification for application, ratings and performance.
Part 2: 1993	Specification for engines.
Part 3:1993	Specification for alternating current generators for generating sets.
Part 4:1993	Specification for controlgear and switchgear.
Part 5:1993	Specification for generating sets.
BS EN 55014: 1997	Electromagnetic compatibility.
BS EN 60095	Lead-acid starter batteries
Part 1:1993	General requirements and methods of test.
Part 2:1993	Dimensions of batteries and dimensions and marking of terminal.
BS EN 60623:1996	Vented nickel-cadmium prismatic rechargeable single cells
BS EN 61056	Portable lead-acid cells and batteries (valve regulated types)
Part 1:1993	General requirements, functional characteristics. Methods of test.

Other references/information sources

BS EN 60034–1:1995	Rotating electrical machines. Rating and performance.
BS EN 60034–3:1996	Rotating electrical machines. Specific requirements for turbine-type synchronous machines.
CIBSE Guide B 10	Electrical power
DEO(W)/DEO TB 59/94:1994	Health and safety – supply voltage – harmonisation throughout the EU – its impact on the MoD estate
DEO(W)/DEO TB 96/18:1996	MoD electricity safety rules and procedures
DEO(W)/TSO SRP 01:1996	Electricity
DEO(W)/DEO TB 97/08:1997	Issue of MoD electricity safety rules and procedures SRP 01

DEO Technical publications index: June 1997.

Electricity at Work Regulations 1989

Electricity Association Engineering Recommendations:

G 5/4:	Limits for harmonics in the UK electricity supply system.
G 59:	Parallel running.

Electrical Contractors Association (ECA) Guidance

HSE Approved Code of Practice: Provision and Use of Work Equipment Regulations 1998.

HVCA Standard Maintenance Specification for Mechanical Services in Buildings – Volumes 5: Electrics in buildings (1990–92)

IEE	Code of practice for in-service inspection and testing of electrical equipment

IEE publication: BS 7671:1992: Requirements for electrical installation – IEE wiring regulations; sixteenth edition

Provision and Use of Work Equipment Regulations 1998

PSA/DEO SSME 1	Electrical installations 1985
PSA/DEO MEEG 3/21	Prime movers, generators & PSA power supplies 1983.
PSA/DEO MEEG 3/22	Transmission and distribution 1983
PSA/DEO MEEG 3/23	LV electrical installations and equipment 1986
PSA/DEO MEEG 3/24	LV Installations in hazardous areas 1983
PSA/DEO MEEG 6/22	Transmission and distribution 1986

STANDBY POWER GENERATING SETS

	DIESEL ENGINE		
YEARS	DESCRIPTION	INSPECTION	MAINTENANCE
30	Turbo-charged diesel engine with 4 to 16 cylinders, to BS 5514 and BS 7698 with power output from 50 kVA to 2000 kVA and with a speed of 1500 rpm. Engine block of cast iron, with jacket water heater and with crankshaft and connecting rods of forged steel. Engine to have hours run metering and to be CE marked.	*Weekly or after 10 hours running on standby duty:* from cold and before starting engine, check fuel oil, lubrication oil and coolant levels. Check fule solenoid. While running, continue visual and aural checks, check ventilation. Complete weekly visit record, report defects.	Those operating and maintaining standby generators are to be 'competent persons' as required by Regulations 6 & 9 of the 'Provision and Use of Work Equipment Regulations 1998'. Maintenance to be strictly in accordance with manufacturer's requirements and/or HVCA/CIBSE Standard Maintenance Specification. Records of maintenance & hours run to be kept.
20	Naturally aspirated diesel automotive or traction type engine with 3, 4 or 6 cylinders, to BS 5514, with power output 25 kVA to 100 kVA and speed in range 1500 rpm to 3000 rpm. Engine to have hours run metering and to be CE marked.	*Six monthly or after 100 hours running on standby duty:* before starting engine check specific gravity of coolant and fuel tank breather. Check function of relays, switchgear, contactors, fuses and tripping devices, alarms and shutdown actuators. Check fuel, exhaust and air pipes and check anti-vibration mounts. While running, continue visual and aural checks. Complete records.	*Weekly or after 10 hours running on standby duty:* run generator set for 5 to 10 minutes off load.

(Note: some organisations specify 2–4 weekly testing, e.g. 30 minutes off load and 4 hours on load).

Three monthly: greasing of re-greasable items. |
| U | Unclassified i.e. diesel engines not manufactured/conforming to the above standards. | *Annually or after 1500 hours running on standby duty:* check V belts and tension, hoses and filters; renew as necessary. Check for condensed water deposits in fuel and lubricating oil systems. Check operation of coolant heater and for defects due to vibration. While running continue visual and aural checks. Complete records. | *Six monthly or after 100 hours running on standby duty:* simulate mains failure and check start-up and load acceptance. Run for four hours on full load or at not less than 75% full load.

Annually or after 1500 hours running on standby duty: change/clean filters, change lubricating oil, drain sediment from fuel tank. Arrange black start.

Two to four yearly (depending on hours run): replace belts and hoses. Overhaul to manufacturer's specification. |

Adjustment Factors

Manufacturer not approved to standard BS EN ISO 9001: –5 years

Assumptions – Design & Installation

Installation to be in strict accordance with manufacturer's instructions and industry good practice (e.g. BSRIA/CIBSE/ECA/HVCA guidance and relevant MoD standards).

Installers to be trained and qualified by the supplier and to be fully conversant with requirements for the engine and alternator.

Full information relating to site conditions and the working environment must be provided by the customer to the manufacturer at design stage, as indicated within BS 5514. Information relating to engine performance and site facilities must be provided by the manufacturer to the customer. These factors will affect post installation performance.

Maintenance specification to be prepared by the manufacturer specifically for client's installation and operating environment.

Hours run metering to be available, on which to base predictive maintenance.

Engine to be connected to chassis via anti-vibration mounts.

Speed governing to BS 5514.

A safety device is required to protect the engine against the effects of over-speed if the governor fails.

High altitude, high temperature and high humidity environments reduce engine performance. De-rating factors should be applied.

Jacket water heater required to maintain satisfactory coolant temperature and to prevent condensation in cylinders and injectors.

Stainless steel or other corrosion resistant bellows required to connect engine exhaust to fixed exhaust outlet.

Engine room fire extinguishing system should be carbon dioxide or aqueous film forming foam (AFFF) to minimise damage to aspiration systems and injectors. Dry powder is an effective extinguisher but can cause considerable damage to plant.

Fire protection system to engine room/enclosure should be linked to engine control panel to initiate engine shut down, fuel dumping from local storage tank, engine shut down alarm and closure of ventilation dampers.

Maintenance and performance records should be retained for the life of the plant.

Assumptions – Commissioning

Commissioning to be in strict accordance with manufacturer's instructions and relevant BSRIA/CIBSE/ECA/HVCA guidance and relevant MoD standards.

Testing of engine shutdown feature and of fire protection system is essential.

Acceptance of installation in the work place to be in accordance with Provision and Use of Work Equipment Regulations.

Maintenance work to be carried out by suitably qualified and experienced personnel.

Key failure modes

Battery failure.

Low oil pressure*

High oil temperature*

High coolant temperature*

Low coolant level*

Engine overspeed*

*(Failure modes marked * to precipitate automatic shutdown and fuel isolation.)*

Key durability issues

Standard of manufacture and quality of materials.

Quality of handling, installation and commissioning.

Site environmental conditions.

Quality of maintenance & inspection regimes.

Site services to manufacturer's requirements.

Provision of jacket water heater.

Maintenance requirements should be based on elapsed time and hours run.

Notes

The inspection and maintenance information contained herein is for guidance only. Manufacturer's and professional adviser's recommendations to take precedence.

Warning indicators on the control panel should reflect the failure modes.

Weekly running of engine off-load is not considered good practice, and can cause build up of soot on injectors. Checking and re-charging starter battery is of key importance. Six monthly full load run will clear carbon deposits from injectors.

Notes – Generator set overall

Generator to be suitable for continuous running at full load for unlimited periods with a variable load. It must also be capable of withstanding an overload of 110% for a period of one hour in any twelve hour period to requirements of BS 5514.

Generator control systems to BS 7698 and to be able to withstand electrostatic discharge and radiated emissions to standard of IEC 800 series and not to interfere with adjacent systems.

Warning signals and engine shutdown limits to be specified at design stage.

Supplier / installer of generator to be fully conversant with connection and installation requirements and shutdown limits of both engine and alternator manufacturers.

Customer to be conversant with the operating and servicing requirements for engine, alternator, control panel and starter battery.

STANDBY POWER GENERATING SETS *(continued)*

	ALTERNATOR		
YEARS	*DESCRIPTION*	*INSPECTION*	*MAINTENANCE*
30	Brushless alternator of steel and cast iron to BS 4999, BS 5000 and BS 7698. Two bearing construction and with flexible coupling at drive end. Enclosure to IP 22 or better with speed in range 1000 rpm to 1500 rpm, having insulation Class H or better with three phase and > 100 kVA output, complete with underfrequency protection, winding thermistors and stator heater.	To be integrated with engine and battery inspections. *Weekly:* Check vents for obstructions. While running, carry out visual and aural checks. *Six monthly:* check function of relays, switchgear, contactors, fuses and tripping devices, alarms and shutdown actuators. While running continue visual and aural checks. *Annually:* Check 'black start' capability, Check bearing condition by vibration monitoring.	To be integrated with engine maintenance requirements. *Six monthly:* clean internally with compressed air. Simulate mains failure and check start-up, load acceptance and stability. *Annually:* grease re-greasable bearings.
25	Brushless alternator of die cast aluminium frame or steel and iron with aluminium components, and to BS 4999 and BS 5000. Single bearing construction and close coupled. Enclosure to IP 21 with speed > 1500 rpm having insulation Class F or better, single or three phase output < 100 kVA and with underfrequency protection.		
U	Unclassified i.e. alternator not manufactured/ conforming to the above standards		

Adjustment Factors

Manufacturer not approved to standard ISO 9001: –5 years

Assumptions – Design & Installation

Installation to be in strict accordance with manufacturer's instructions and industry good practice (e.g. BSRIA/CIBSE/ECA/HVCA guidance and relevant MoD standards).

Installers to be trained and qualified by the supplier and be fully conversant with requirements for engine and alternator.

Full information relating to site conditions and the working environment (e.g. temperature and humidity range) to be provided by the customer to the manufacturer at the design stage, in accordance with BS 5514.

Hours run metering to be available, on which to base predictive maintenance.

Requirements for the alternator to have one or two bearing shaft support, to have close or flexible coupling and the use of anti-vibration mounts to be discussed between the client's professional advisor and the manufacturer at design stage.

Protection is required to prevent failure modes such as electrical overload, short circuit, earth leakage current and reverse power. Differential protection on large sets (2000kW and over) where stator winding connectors are brought out separately to the star point.

Electrical integrity of alternator and power system to be protected against over-current, short circuit, over/under voltage, under frequency and overload. Suitable alarms and shutdown parameters to be specified at design stage.

Temperature detectors required in stator windings and shaft bearings to indicate excesses of temperature.

Adequate fan cooling required to maximise heat transfer and minimise hot spot differentials.

Permanent magnet excitation required to overcome loss of residual excitation.

Bearings to be re-greasable.

Rotating diode failure indicator is required for larger machines

Assumptions – Commissioning

Commissioning to be in strict accordance with manufacturer's instructions and relevant BSRIA/CIBSE/ECA/HVCA guidance and relevant MoD standards.

Key failure modes

Stator insulation faults.
Over/under voltage.
Electrical overload.
Rotor failure.
Excitation system failure.
Earth leakage.
Short circuit.
Failure of automatic voltage regulator (AVR).

Key durability issues

Standard of manufacture and quality of materials.

Quality of handling, installation and commissioning.

Site environmental conditions.

Quality of maintenance and inspection regimes.

Site services to manufacturer's requirements.

Overloading of alternator at start-up.

Maintenance based on elapsed time and hours run.

Notes

Inspection and maintenance information contained herein is for guidance only. Manufacturer's and professional adviser's recommendations to take precedence.

Overloads in excess of 110% for more than one hour in twelve will lead to overheating of alternator windings and a shortening of life expectancy. See BS 5514 for further details. The alternator will normally be rated continuously for up to 110% overload.

Ingress protection index of IP 22 is standard. Higher values can be achieved to improve protection against dust and spray but may result in application of a de-rating factor. Check with manufacturer.

Inlet and outlet air filters can be fitted to alternators but de-rating factors of 5% apply to each.

Recommended spares list to include spare rotating diode(s), spare rotating diode fuses and spare fuses for automatic voltage regulator.

Maintenance record should be retained for the life of the plant.

Differential fault protection should be considered for low voltage alternators rated at 2MW and over. Sets of current transformers (CTs) wired back to back from stator and adjacent LV switchgear sense faults between the sets of CTs.

Notes – Generator set overall

Generator to be suitable for continuous running at full load for unlimited periods with a variable load. It must also be capable of withstanding an overload of 110% for a period of one hour in any twelve hour period to requirements of BS 5514.

Generator control systems to BS 7698 and to be able to withstand electrostatic discharge and radiated emissions to standard of IEC 800 series and not to interfere with adjacent systems.

Warning signals and engine shutdown limits to be specified at design stage.

Supplier / installer of generator to be fully conversant with connection and installation requirements and shutdown limits of both engine and alternator manufacturers.

Customer to be conversant with the operating and servicing requirements for engine, alternator, control panel and starter battery.

STANDBY POWER GENERATING SETS *(continued)*

STARTER BATTERIES

YEARS	DESCRIPTION	INSPECTION	MAINTENANCE
25	High performance plant-batteries to BS 6290.	*Six monthly:* Inspect cells. Check connections for security and corrosion. Check electrolyte levels and adjust. For larger batteries, take readings on 10% of cells and analyse viz. specific gravity, voltage, temperature and electrolyte level.	*Six monthly:* Clean cells. Clean terminals, dry and apply silicone grease. *Yearly:* carry out discharge check.
25	Pocket plate nickel cadmium batteries to BS EN 60623.	*Yearly:* Inspect cells. Check connections for security and corrosion. Check electrolyte levels and adjust. For larger batteries, take readings on 10% of cells and analyse viz. specific gravity, voltage, temperature and electrolyte level.	*Yearly:* Clean cells. Clean terminals, dry and apply silicone grease. *Yearly:* carry out discharge check.
20	Free electrolyte nickel cadmium batteries to BS EN 60623.	*Yearly:* Inspect cells. Check connections for security and corrosion. Check electrolyte levels and adjust. For larger batteries, take readings on 10% of cells and analyse viz. specific gravity, voltage, temperature and electrolyte level.	*Yearly:* Clean cells. Clean terminals, dry and apply silicone grease. *Yearly:* carry out discharge check.
15	Valve regulated sealed re-chargeable batteries to BS 6290 & BS EN 61056.	*Three monthly:* Inspect cells. Check connections for security and corrosion.	*Three monthly:* Clean cells. Clean terminals, dry and apply silicone grease. *Yearly:* carry out discharge check.
15	Vented-type stationary lead-acid batteries to BS 6290 & BS EN 61056.	*Three monthly:* Inspect cells. Check connections for security and corrosion. Check electrolyte levels and adjust. For larger batteries, take readings on 10% of cells and analyse viz. specific gravity, voltage, temperature and electrolyte level. *Six monthly:* for larger batteries take readings as above for all cells.	*Three monthly:* Clean cells. Clean terminals, dry and apply silicone grease. *Yearly:* carry out discharge check.
5	Lead-acid automotive batteries to BS EN 60095.	*Three monthly:* Inspect cells. Check connections for security and corrosion.	*Three monthly:* Clean cells. Clean terminals, dry and apply silicone grease. *Yearly:* carry out discharge check.
U	Unclassified i.e. batteries not manufactured/conforming to the above standards.	As above.	As above.

Adjustment Factors

Surrounding temperature in excess of 25°C for prolonged periods: −5 years

Assumptions Design & Installation

Installation to be in strict accordance with manufacturer's instructions and industry good practice (e.g. BSRIA/CIBSE/ECA/HVCA guidance and relevant MoD standards).

Personnel working with batteries to be trained to work on live equipment and to be fully conversant with the special safety requirements.

Manufacturer's instructions relating to pre-installation delivery and storage must be complied with.

If storage is extended, then batteries are to be recharged according to manufacturer's instructions every three to four months depending on storage ambient temperatures.

Effective temperature control is critical to the longevity of batteries. Temperature control to be provided where necessary to ensure that surrounding temperature does not exceed 25°C for prolonged periods.

Battery voltage and polarity to be checked before connection.

Battery terminals to be lightly greased with silicon according to manufacturer's requirements.

Special instructions are to observed for connection of parallel sets of batteries.

Battery voltage to be checked against the specification on completion.

Batteries to be installed as close as possible to the engine to minimise voltage drop.

Batteries to be installed on vibration free mount and not on engine chassis.

Where electrolyte topping-up is permitted/required only electrolyte supplied by the battery manufacturer is to be used.

Battery charger maintenance to manufacturer's requirements or to ECA/CIBSE Standard Maintenance Specification.

Assumptions – Commissioning

Commissioning to be in strict accordance with manufacturer's instructions and relevant BSRIA/CIBSE/ECA/HVCA guidance and relevant MoD standards.

Pre-commissioning charge routine to be carried out according to manufacturer's instructions.

Ensure battery charger is matched to the charging characteristics of the battery.

Key failure modes

Prolonged storage (without maintenance attention) before initial use.

Engine fault requires unacceptable number of battery starts.

Mis-interpretation of polarity during connection.

Foreign body in cell.

Corrosion of terminals.

Short circuiting of terminals.

Leakage.

Lack of maintenance, e.g. cells not topped up (where topping up is a requirement); loose/dry terminal connection; lack of charge/discharge cycle.

Key Durability Issues

Standard of manufacture and quality of materials.

Quality of handling, installation and commissioning.

Site environmental conditions.

Frequency of maintenance. Irregular maintenance is the most common cause of standby generator battery failure.

Battery conditioning cycles.

Ambient temperature. Battery life will be diminished in direct proportion to the length of time that the surrounding temperature exceeds 25°C. For every 10 °C rise in ambient temperature above 20 °C, the life expectancy of a battery is reduced by approximately 25&50%.

Notes

Maintenance information is provided for guidance only: specific manufacturer's requirements take precedence.

All the battery types listed benefit from periodic conditioning, charging and discharging, according to manufacturer's requirements.

Lead-acid cells benefit from regular use in service.

Maintenance requirements for un-sealed lead-acid cells are contained in ECA/CIBSE Standard Maintenance Specification.

The EC is to propose that EU member states should collect and re-cycle all batteries by 2002, if directive is adopted. The draft directive also proposes phasing out of cadmium use by 2008.

Valve regulated and sealed re-chargeable batteries are low maintenance due to gas recombination process.

Notes – Generator set overall

Generator to be suitable for continuous running at full load for unlimited periods with a variable load. It must also be capable of withstanding an overload of 110% for a period of one hour in any twelve hour period to requirements of BS 5514.

Generator control systems to BS 7698 and to be able to withstand electrostatic discharge and radiated emissions to standard of IEC 800 series and not to interfere with adjacent systems.

Warning signals and engine shutdown limits to be specified at design stage.

Supplier / installer of generator to be fully conversant with connection and installation requirements and shutdown limits of both engine and alternator manufacturers.

Customer to be conversant with the operating and servicing requirements for engine, alternator, control panel and starter battery.

LOW VOLTAGE SWITCHGEAR

Scope

This section provides data on switchgear and switchboard assemblies employed as part of low voltage (less than 1000V) distribution systems in commercial or industrial premises. It includes data on circuit breakers, fuse switches, contactors and switchboard assemblies. Motor control centres, metering, systems monitoring and electrical protective devices are excluded from the scope of this study, as are installations in potentially hazardous areas.

Standards cited

BS 88	Cartridge fuses for voltages up to and including 1,000V ac and 1,500V dc
Part 2:1988	Specification for fuses for use by authorised persons (mainly for industrial application). Section 2.2: Additional requirements for fuses with fuse-links for bolted connections.
BS 4293:1983	Specification for residual current-operated circuit breakers.
BS 5486	Low-voltage switchgear and controlgear assemblies.
Part 11:1989	Specification for particular requirements of fuseboards.
BS 7430:1998	Code of practice for earthing.
BS EN 60269	Low-voltage fuses
Part 1:1999	General requirements. Also numbered BS 88: Part 2: Section 2.1.
Part 2: 1995	Supplementary requirements for fuses for use by authorised persons (fuses mainly for industrial application).
BS EN 60439	Low-voltage switchgear and controlgear assemblies.
Part 1:1999	Type-tested and partially type-tested assemblies.
Part 3:1991	Particular requirements for low-voltage switchgear and control-gear assemblies intended to be installed in places where unskilled persons have access to their use – distribution boards.
BS EN 60529:1992	Degrees of protection for enclosures.
BS EN 60898:1991	Specification for circuit breakers for overcurrent protection for household and similar installations.
BS EN 60947	Specification for low-voltage switchgear and controlgear
Part 1:1999	General rules.
Part 2:1996	Circuit breakers
Part 3:1999	Switches, disconnectors, switch-disconnectors and fuse combination units.
Part 4: (various)	Contactors and motor-starters. Section 1: Electromechanical contactors and motor-starters.
BS EN 61000: (various)	Electromagnetic compatibility (EMC)
BS EN 61008	Residual current operated circuit breakers without integral overcurrent protection. For households and similar uses (RCCBs)
Part 1:1995	General rules.
Part 2–1:1995	Applicability of the general rules to RCCBs functionally independent of line voltage.

Other references / Information Sources

BRE	Magnetic fields and building services.
BS 7671:1992 IEE	Wiring Regulations 16th Edition. Requirements for electrical installations.
CIBSE / ECA / HVCA	Standard maintenance specification for services in buildings. Volume 5, Electrics in buildings.
DEO	Technical publications index: June 1997.
EIEMA	Guide to residual current devices.
EIEMA	Guide to type tested and partially type tested assemblies; low voltage switchgear and controlgear assemblies: BS EN 60439–1.
EIEMA	Guide to forms of separation; low voltage switchgear and controlgear assemblies: BS EN 60439–1.
PSA/DEO MEEG 3/23	LV electrical installations and equipment.
PSA/DEO MEEG 6/23	Maintenance of LV electrical installations and equipment.
Health & Safety at Work Etc Act 1974	
The Electricity at Work Regulations 1989.	
The Workplace (Health, Safety and Welfare) Regulations 1992.	
The Electrical Equipment (Safety) Regulations 1994 – The Low Voltage Directive.	
Provision and Use of Work Equipment Regulations 1998	
DEO SRP01	Electricity: MoD electricity safety rules and procedures.

LOW VOLTAGE SWITCHGEAR

YEARS	DESCRIPTION	INSPECTION	MAINTENANCE
25	Moulded case circuit breaker (MCCB) to BS EN 60947 16A to 200A.	**Annually** – Visual check of condition. Check for signs of overheating. Check for loose connections.	**Annually** – Check operation and free movement. Replace if damaged. Clean visible surfaces with damp soapy cloth while isolated, avoid solvents and abrasive cleaners.
25	Moulded case circuit breaker (MCCB) to BS EN 60947 100A to 1600A.	**Annually** – Visual check of condition. Check for signs of overheating. Check for loose connections.	**Annually** – Check operation and free movement. Open breaker, check and replace components to manufacturer's instructions.
20	Miniature circuit breaker (MCB) to BS EN 60898 / BS EN 60947 with optional 'tripped' indicator; residual current device (RCD) to BS EN 61008 / BS 4293 & residual current operated circuit breaker with integral overcurrent protection (RCBO) to BS EN 61009. Breaker to be 1,2,3 or 4 poles and have operating current ≤125A.	**Annually** – Visual check of condition. Check for signs of overheating. Check for loose connections.	**Annually** – Check operation and free movement. Replace if damaged. Clean visible surfaces with damp soapy cloth while isolated, avoid solvents and abrasive cleaners.
25	Switchgear in powder coated sheet steel enclosure to BS EN 60947 with rating ≤800A. Fuses to BS 88 & BS EN 60269.	**Annually** – check contacts and free movement of moving parts, check for evidence of arcing, note condition and characteristics of load. Check for signs of overheating. Check for loose connections. Check rating of fuses fitted.	**Annually** – operate switch, clean and re-lubricate as appropriate, check and replace components to manufacturer's instructions.
20	Switchgear in powder coated sheet steel enclosure to BS EN 60947 with rating ≤100A. Fuses to BS 88 & BS EN 60269.		
25	Contactors, glass-reinforced polyester or die cast light alloy 3 or 10 pole rated 10A to 800A to BS EN 60947.	**Annually** – Visual check of condition. Check for signs of overheating. Check for loose connections.	**Annually** – Check operation and free movement. Open breaker, check and replace components to manufacturer's instructions.
25	Air circuit breakers rated from 800A to 4,000A	**Annually** – Visual check of condition. Check for signs of overheating. Check for loose connections.	**Annually** – check operating mechanism, contacts & arcing chambers, check & replace components to manufacturer's instructions.
25	Switchboard or switchgear assemblies constructed from 2mm powder coated / enamelled sheet steel or equal selected for environment concerned, type tested to BS EN 60439 and ASTA certified.	**Annually** – Check for signs of corrosion and damage, overheating and loose connections particularly for main cables and busbars. Check balance of 3 phase load on panel. Check instrumentation and protective devices. Check indicators, neutral and earth connections and continuity. Check for moisture and accumulation of dirt. Check availability and use warning notices and locks. Keep records of work for life of equipment. Check switches labels and designation of switches.	**Annually** – Tighten bolted connections with great care, particularly after several cycles of maintenance. Check connection impedances. Redistribute unbalanced loads if possible. Replace or repair defective instruments. Replace failed indicator lamps. Deal with sources of moisture, dirt and corrosion. Clean all accessible surfaces.
U	Unclassified i.e. switchgear of unknown materials not manufactured to the above standards.	As above	As above

Note: All inspection and maintenance to be carried out in accordance with statutory safety rules and procedures as set out in the Electricity at Work Regulations and the Provision and Use of Work Equipment Regulations.

Adjustment factors

Maintenance not to requirements of manufacturer or professional adviser: –10 years.

Assemblies and equipment not type tested, fully or partially, to BS EN 60439: –10 years.

Dusty or polluted environment: –5 years.

Note: The above factors are not cumulative; the largest applicable factor should be applied.

Assumptions – Design & Manufacture

The lives of switchgear will vary according to the resistive and inductive component of the electrical load. Load characteristics to be assessed at design stage.

Manufacture to be approved to BS EN ISO 9001 or BS EN ISO 9002.

Manufacturer to be a member of an accredited UK trade association or regulating body (such as the EIEMA Low Voltage Switchboard Division).

Assemblies and equipment to be type tested, fully or partially, to BS EN 60439.

All products to be CE marked to indicate compliance with the Low Voltage Directive. All products to comply with the Electromagnetic Compatibility Directive.

Electrical equipment earthing and sizing of earthing conductors to BS 7430.

Forms of separation and types of insulation to be defined at design stage according to operating requirements. BS EN 60439 provides performance criteria.

The minimum ingress protection index to be IP2X to BS EN 60529.

Where equipment has been held in store prior to installation, the packaging and contents are to be checked for signs of corrosion.

Tinning and silver plating of switch contacts and conductor connections can prolong life and reduces the amount of copper required to achieve a rating.

All equipment to be Association of Short-circuit Testing Authorities (ASTA) certified.

Rating of switchgear to be reduced as ambient temperature increases.

Fuse boards to BS 5486 Part 11.

Cartridge fuses to BS EN 60269 and BS 88 Part 2 Section 2.2.

Enclosures and cabinets to be adequately ventilated and constructed to avoid ingress of dirt and formation of condensation.

BS EN 60947 requires ambient air temperature to be {40°C and {35°C mean over 24 h. For every 10°C rise in ambient temperature above 30°C, life of contacts can be reduced by approximately 50%.

Where local temperatures within or around switchgear can exceed 50°C environmental control to be provided or switchgear to be relocated.

In low temperature situations, panel heaters required.

Enclosures for use in damp/humid environments to have additional external protective coating. Enclosures for use externally to have rain canopy and to be of appropriate IP rating to BS EN 60529.

Assumptions – Installation & Commissioning

Transport and storage requirements, humidity and temperature limits to BS EN 60947.

Installation in strict accordance with manufacturer's instructions and BS EN 60947.

Installation to be inspected and tested, including functional tests on completion; safe means of access and working space to be assessed all in accordance with BS 7671 (IEE wiring regulations). Copies of all certificates to be retained on site.

Copies of factory test certificates to be made available and retained on site.

Pre-energisation testing to include insulation and continuity testing – Ducter (low resistance) testing for busbar joints.

Electrical changeover systems, switches, contactors, breakers, interlocks, panel supplies to be checked and operated after connection of primary supplies.

Commissioning test records to be tabulated and retained on site.

Key failure modes

Corrosion due to lack of environmental control during storage or in situ.

Damage due to high temperatures within enclosures.

Overheating of joints /connections due to harmonics drawn by load.

Tracking due to excessive deposits of dust or condensation.

Post maintenance fault.

Loosening of bolted connections over time due to vibration.

Incorrect switching procedure by operator.

Failure of protection devices.

Overcurrent.

Short circuit between phases or to neutral or earth.

Change of load profile and/or load characteristics; increase of load beyond forecast limits.

Key durability issues

Protection against moisture and high temperatures.

Correct specification, i.e. matching of switchgear to load.

Electrical endurance viz. contact life, depends on percentage of full load being switched, the frequency of operation and the utilisation category, e.g. AC–1 predominantly resistive and AC–4 predominantly inductive (ref. BS EN 60947–1).

Tinning or silver plating of contacts and points of contact helps to reduce contact impedance, heat generation and requirement for heat dissipation.

Fuse switches should replace switch fuses over 100A rating.

Notes

Loose switchgear can be satisfactorily contained in enclosures constructed from any of the following materials: cast aluminium, polycarbonate, polyester, powder coated sheet steel, powder coated sheet steel, anodised aluminium panels, plated sheet steel and stainless steel. Choice depends on environment, degree of protection required and function.

Empty enclosures fitted with switchgear by others are not required to be CE marked under EMC legislation, or BS EN 61000 series.

Compliance with Electrical Equipment (Safety) Regulations 1994 (the low voltage directive) is mandatory. Low voltage assemblies to BS EN 60947 comply with the directive.

Persons who operate and maintain switchgear, and duty holders, to be competent and to employ safe systems of work (see Regulations 6 and 9 of the Provision and Use of Work Equipment Regulations 1998).

Maintenance of electrical systems, e.g. protection devices to prevent danger, is mandatory under the Electricity at Work Regulations.

Inspection and maintenance to be carried out only by experienced and qualified personnel under the control of a duty holder as defined in the Electricity at Work Regulations. Frequencies to be agreed with professional adviser in accordance with BS 7671.

Frequency of inspection and maintenance to be based on manufacturer's recommendations and a knowledge of the particular load characteristics – ratio of resistive and inductive components, percentage of load being switched and frequency of operation.

The rating of MCBs and contactors above which maintenance is practicable varies with manufacturer.

Switchgear utilisation categories for a full range of applications are contained in Annex A of BS EN 60947–1.

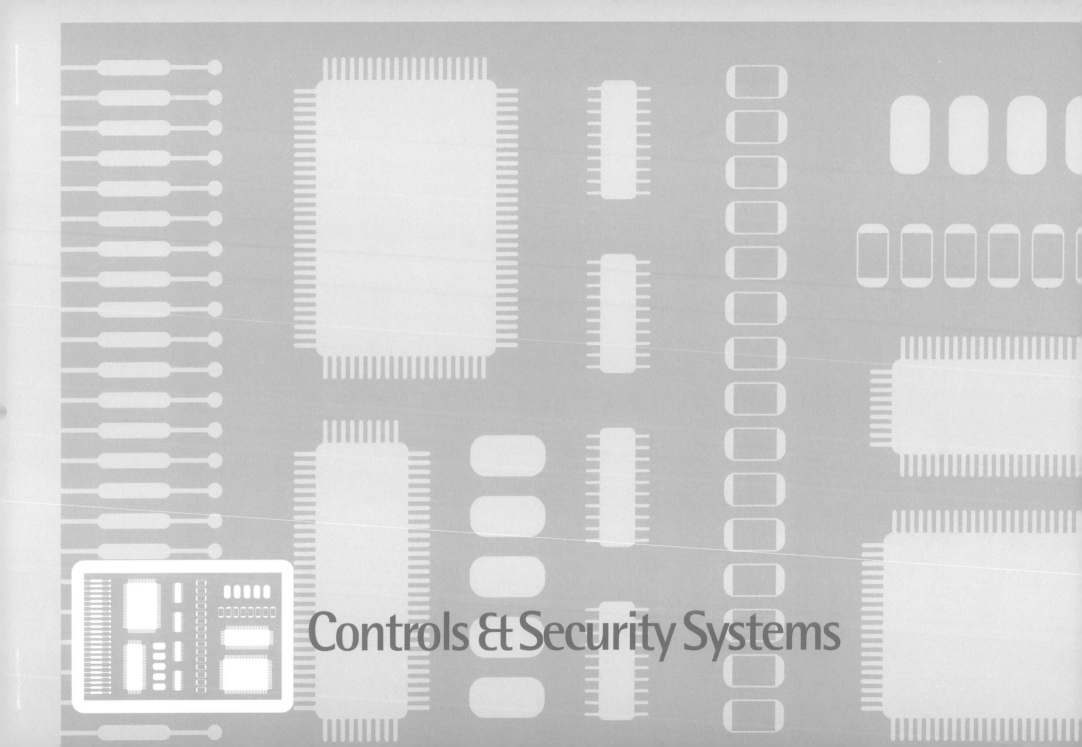

Controls & Security Systems

CONTROL SYSTEMS

Scope

This section provides data on control system outstations and standalone controls used in commercial and light industrial applications. It includes only the main hardware components, i.e. the protective enclosure, printed circuit board, transformer, back-up battery, terminals, LV wiring and temperature sensor. Software systems are excluded, as are outstations associated with fire and security systems. Outstations may operate in isolation or be linked to a central station on large or complex sites.

The following component sub-type is included within this section:

Component Sub-type	Page
Outstation and standalone controls hardware	106

Standards cited

BS 3535	Isolating transformers and safety isolating transformers
Part 1:1990	General requirements.
Part 2:1990	Specification for transformers for reduced system voltage.
BS 6221	Printed wiring boards (various parts).
BS EN 60249	Base material for printed circuits
Part 1:1993	Test methods
BS EN 60529: 1992	Specification for degrees of protection provided by enclosures (IP code).
prEN 13646:1999	Building control systems – equipment characteristics.
prEN ISO 16484	Building control systems
Part 1:1999	Overview and definitions.
Part 2:1999	HVAC control system functionality

Note: There is not (at the time of writing) any standards specifically for a complete control system outstation.

Other references / information sources

BS 3938: 1973	Specification for current transformers.
BS EN 60068	Environmental testing
Part 1:1995	General and guidance
Part 2	Test methods (various parts)
BS EN 60439	Specification for low voltage switchgear and controlgear assemblies.
Part 1: 1994	Specification for type-tested and partially type-tested assemblies.
BS EN 60721	Classification of environmental conditions
Part 1:1996	Environmental parameters and their severities
BS EN 60742: 1996	Isolating transformers and safety isolating transformers. Requirements.
BS EN 60947	Specification for low-voltage switchgear and control gear.
Part 1:1999	General rules.
BSRIA AG 2/94: 1994	BEMS performance testing.
BSRIA BR 64: 1985	Performance of HVAC systems and controls in building.
BSRIA IP 6/85:1985	Selection of building energy management system.
BSRIA AH 1/90: 1990	Guide to BEMS Centre Standard, Volume 1.
BSRIA AH 1/90: 1990	Guide to BEMS Centre Standard, Volume 2.
BSRIA AH 1/92: 1992	Commissioning of BEMS – a code of practice.
IEE	Code of practice for in-service inspection and testing of electrical equipment
NHS	NHS Model Engineering Specification C54: Building Management System.
PSA/DEO MEEG 2/12	Computer based building management systems (1987)
PSA/DEO MEEG 5/12	Computer based building management systems (1987)
Defence Estates Standard Specification (M&E) No. 015.	Building Management Systems (formerly PSA specification no.15).

	OUTSTATION AND STANDALONE CONTROLS HARDWARE		
YEARS	DESCRIPTION	INSPECTION	MAINTENANCE
15	Outstation / standalone controller hardware to draft standards prEN 13646 and prEN ISO 16484. Enclosure protection to BS EN 60529, IP 41 or IP 54 as appropriate. Transformers to BS 3535. 'Plug in' printed circuit boards to BS 6221 and BS EN 60249–1 (i.e. for use with motherboard). Manufacturing process quality assured to BS EN ISO 9001.	Inspection and testing in accordance with requirements of IEE Code of practice for in-service inspection and testing of electrical equipment. **Annually:** Outstations and control units should be subjected to periodic visual inspection to check security, integrity and for damage, particularly to door seals and security of incoming cables. Check environmental conditions are within prescribed limits. Carry out voltage check on all power supplies. Check stand-by batteries/uninterrupted power supplies (UPS) against manufacturer's specification.	Maintenance in accordance with manufacturer's recommendations and HVCA Maintenance Specifications – Volume III, to include (typically): **Annually:** If necessary, clean to remove dust. Particularly dusty environments may require a circulation fan and filter for the outstation enclosures. It is recommended that 're-commissioning' be carried out periodically, to include circuit and component testing. Batteries should be changed approximately every three years.
10	Outstation / standalone controller hardware to draft standards prEN 13646 and prEN ISO 16484. Enclosure protection to BS EN 60529, IP 41 or IP 54 as appropriate. Transformers to BS 3535. Printed circuit boards to BS 6221 and BS EN 60249–1. Manufacturing process quality assured to BS EN ISO 9001.		
5	Outstation / standalone controller hardware to draft standards prEN 13646 and prEN ISO 16484. Enclosure protection to BS EN 60529, IP 41 or IP 54 as appropriate. Transformers to BS 3535. Printed circuit boards to BS 6221 and BS EN 60249–1. Manufacturing process not quality assured to BS EN ISO 9001.		
15	Unclassified i.e. control systems hardware not formed or tested to the above standards.		

Adjustment factors

Changes in technology, amount of plant, information requirements etc often result in control hardware being changed before the useful life has expired. The use of 'motherboards' with 'plug-in circuit boards' may help to extend the life of the control hardware. +5 years.

Assumptions – Design & Installation

Design and installation of controls system in strict accordance with manufacturer's instructions and industry good practice (e.g. BSRIA/CIBSE/HVCA/IEE guidance and relevant MoD standards).

Adequate instrumentation should be fitted to enable accurate commissioning and to provide accurate readings for future performance monitoring.

The degree of protection provided by the equipment enclosure may be reduced to BS EN 60529 category IP41 when the enclosure is fitted within another panel such as a motor control centre. Otherwise, enclosures should be constructed to give minimum IP54 protection.

Control system design to be appropriate for intended environment. Draft standard prEN 13646 defines seven environmental categories.

Outstations to be located within protective metal cabinets with a lockable door.

Operating instructions and diagrams to be provided within the metal cabinet.

Any cables (for power, plant, sensors) to be protected by trunking, protective sleeving or conduit.

Assumptions – Commissioning

Commissioning and testing to be in strict accordance with manufacturer's instructions and industry good practice (e.g. BSRIA/CIBSE/HVCA/IEE guidance and relevant MoD standards).

Commissioning should ideally be carried out by the control system supplier or specialised commissioning firms with experience of the particular type of equipment.

The main plant must be commissioned and operated prior to the controls being commissioned. Once this has been achieved, the controls should be calibrated and adjusted to provide stable operating conditions which match the design specification.

The stability of operation under varying climatic conditions should be checked. At least two post-commissioning visits during the first year are recommended.

Key failure modes

Indoor environment – exposure to high/low temperatures, humidity, particulates and corrosive atmospheres.

Voltage spikes.

Harmonics.

Failure of the control system to accurately control the system , e.g. due to dirty sensing elements; failure to reset controls after they have been overridden; mechanical failure of valve, thermostats etc.

Failures can occur in hot boiler houses, as a result of static electricity and/or or poor soldering.

Poor siting of the units in relation to nearby plant can also cause problems, e.g. vibration and electrical interference.

Key durability issues

Standard of manufacture and quality of materials.

Quality of handling, installation and commissioning.

Poor siting of the units with relation to environment (outstations should be capable of operating within a temperature range of 0°C to 50°C and a relative humidity of 10 to 90% as specified in BSRIA Application Handbook AH 1/90 Standard Specification for BEMS).

Periodic 're-commissioning' can help to identify potential detector failure by measuring their 'signal' or resistance.

FIRE PROTECTION – SPRINKLER SYSTEMS

Scope

This section provides data on fire protection sprinkler systems for use in commercial and light industrial applications. It does not deal with mist or deluge systems, or with smoke detectors or alarms, pressurisation pumps, landing valves or wet and dry risers.

The following component sub-type is included within this section:

Component Sub-type	Page
Glass bulb and fusible link sprinkler heads	111

Standards cited

BS 21:1985	Specification for pipe thread for tubes and fittings where pressure-tight joints are made on the threads
BS 5306 Part 2:1990	Fire extinguisher installations and equipment on premises. Specification for sprinkler systems
BS EN 12259–1:1999	Fixed fire fighting systems. Components for sprinkler and water spray systems.
LPC 1990	Rules for Automatic Sprinkler Installation
LPC 1036:Issue 2: July 1993	Quality schedule for the certification of automatic fire sprinklers
LPS 1039:Issue 3:Oct 1989	Requirements and testing methods for automatic sprinklers
LPS 1048:Issue 3:May 1996	Requirements for certificated sprinkler installers, supervising bodies and supervised installers
LPS 1050:Issue 1:May 1995	Requirements for automatic sprinkler system servicing and maintenance firms
LPS 1213 LPCB	Stockist scheme

Other references/information sources

BRE :1993: BR 225	Aspects of fire precautions in buildings
BSRIA Application Handbook:1992: AH 3/92	Installation, Commissioning and maintenance of fire and security systems
CIBSE Guide E:1997	Fire engineering
DEO(W)/DEO TB 24/93:1993	Fire prevention and fire safety – inspection, testing and maintenance of sprinkler installations
DEO/DEO TB 98/12:1998	Publication of the Crown Fire Standards –
HVCA	Standard Maintenance Specification for Mechanical Services in Buildings: 1990–92: Volume 4 Ancillaries, plumbing and sewerage
Institute of Plumbing	Plumbing engineering services design guide *(withdrawn for revision; due in late 2000)*
	Pressure Systems and Transportable Gas Regulations (1989)
PSA/DEO MEEG 3/28:1983	Fire detection, alarm and extinguishing systems
PSA/DEO MEEG 6/28:1985	Fire detection, alarm and extinguishing systems
PSA/DEO FPG 005:1987	Inspection and testing of sprinkler installations

SPRINKLER HEADS

	GLASS BULB & FUSIBLE LINK SPRINKLER HEADS		
YEARS	DESCRIPTION	INSPECTION	MAINTENANCE
30	Sprinkler heads to BS 5306:Part 2 and Loss Prevention Standard (LPS) 1039. Individual components to BS EN 12259–1. The procurement of sprinkler equipment and full training must be from an LPCB-listed sprinkler equipment manufacture or a certified stockist to LP 1213 or equivalent standard, acceptable to LPCB.	Equipment is subject to requirements of BS 5306:Part 2 and the Loss Prevention Certification Board (LPCB) and examination must be carried out by 'competent persons' as required by the Fire Regulations 1998.	The installer is to provide an inspection and checking programme for the system. Maintain in accordance with manufacturer's requirements, BS 5306:Part 2 , the LPCB requirements and The Fire Regulations 1998.
U	Unclassified i.e. sprinkler heads not to the above standards.	LPS 1048 Certificated installations which are serviced and maintained to LPS 1050 will carry two LPCB Certificates: a LPS 1048 Certificate of Conformity and a LPS 1050 Certificate of Inspection and Servicing. Any 'Review of hazard' must be as referred to in the LPC Rules/BS 5306: Part 2. Sprinklers should also be subjected to visual inspection during commissioning (and then at least annually) of the main components i.e. associated risers, pipework and filters, valves, pressurization units, etc	Automatic sprinkler system servicing and maintenance firms must be certified and registered to LPC 1050. All assembled sprinklers shall be production pressure tested. All assembled glass bulb sprinklers shall be subjected to a test and/or inspection which ensures the identification and rejection of sprinklers with defective bulbs. All finished sprinklers shall be visually inspected in accordance with an inspection list. Sprinklers shall be subjected to additional production tests according to a sampling plan, to include the following: functional test, release temperature tests for glass bulbs and fusible sprinklers, strength of release element and service load measurement.

Adjustment factors

Use in adverse (but not severely corrosive) environments, and without the appropriate corrosion protection: –5 years

Assumptions – Design & Installation

Design and installation of the sprinkler system to be in strict accordance with manufacturer's instructions and industry good practice (eg BSRIA / CIBSE / HVCA / LPC) guidance and relevant MoD standards and following guidance in LPC Rules/BS 5306:Part 2.

Adequate measures should be taken to prevent freezing in any part of the system.

A strainer/filter to be fitted in the supply line to prevent problems with suspended material etc.

Sprinkler guards to be fitted to reduce physical damage to the sprinkler heads.

The sprinkler system must be selected to suit the particular application, i.e. release mechanism, type of discharge, mounting position, operating temperature rating, responsive time, rate of discharge, building type and contents, etc

Provision for testing and measuring shall be provided at the appropriate locations.

Pipework and other components associated with sprinkler systems to have all the necessary supports.

If a sealing compound is used, apply to the sprinkler thread only.

Assumptions – Commissioning

Pre-commissioning cleaning of water systems should be carried out in accordance with good practice (e.g. BSRIA AG2/89 & AG8/91)

Commissioning to be in strict accordance with manufacturer's instructions and relevant BSRIA / CIBSE /HVCA / MoD / LPC other guidance / codes i.e. BSRIA Application Handbook AH/ 3/92 and LPC Rules/BS 5306:Part 2

Key failure modes

Sprinkler heads and systems have very few failure modes. Most 'failures' are associated with human, not mechanical error. These include: valve deliberately shut i.e. shut off too early during fire; shut off to prevent freezing or to carry out maintenance.

Other causes of failures include: inadequate water supply; obstruction to distribution of water; inadequate maintenance; slow operation; defective supply pipework and/or valves; frozen system; damage due to accidental painting or otherwise covering of the sprinkler head.

Corrosion.

Key durability issues

Standard of manufacture and quality of materials.

Quality of handling, installation and commissioning.

Corrosion resistance of materials.

Resistance to physical damage to sprinkler heads.

Notes

Sprinkler durability should not be affected by type of system (dry, alternate, wet), type of release mechanism (solder link or glass bulb), discharge type (conventional, spray, sidewall), mounting position (upright, pendent or horizontal), thread or orifice size or location type (flush, recessed or concealed).

Sprinklers and their components can be manufactured from many materials (stainless steel, copper, copper alloy, brass, chrome, plastic) and wax coating is available for additional corrosion protection.

LPC 1048 will be revised in 2000 and the revision will contain sections of LPC 1050. The revision will also distinguish between the two different types of certified companies i.e. those who design, install and maintain sprinkler systems and those who design and install only.

Transport Systems

ELECTRIC TRACTION LIFTS

Scope

This study provides data on electric traction lifts used for transporting people (in the workplace). Data is provided on the major component parts of bespoke lifts (i.e. lifts for which each component is specified separately), and for 'packaged' lifts which are specified as a complete unit.

It should be noted that the longevity of lift components is a highly complex subject and that for many component parts, replacement will be dictated by statutory inspections. The lives given are therefore indicative only.

The following component sub-types are included within this section:

Component Sub-type	Page
Manufacturers' 'packaged' lifts	119
Bespoke lifts	119
Lift ropes	122
Drive units	124

Standards cited

BS 302		Stranded steel wire ropes
	Part 4:1987	Ropes for lifts
BS 721		Specification for worm gearing
	Part 2:1983	Metric units
BS 4999 (various parts)		General requirements for rotating electrical machines
BS 5000 (various parts)		Rotating electrical machines of particular types or for particular applications
BS 5655		Lifts and service lifts
	Part 5:1989	Specification for the dimensions of standard lift arrangements
	Part 6:1990	Code of practice for selection and installation
	Part 7:1983	Specification for manual control devices, indicators and additional fittings
	Part 8: 1983	Specification for eyebolts for lift suspension
	Part 9:1985	Specification for guide rails
	Part 10	Testing and examination of lifts and service lifts
	Section 10.1	Electric lifts
	Subsection 10.1.1	Commissioning tests for new lifts
	Part 11:1989	Recommendations for the installation of new, and the modernisation of, electric lifts in existing buildings
BS 7255: 1989		Code of practice for safe working on lifts

BS EN 81–1:1998	Safety rules for the construction and installation of lifts – Electric lifts.
BS EN 627:1996	Specification for data logging and monitoring of lifts, escalators and passenger conveyors.
BS EN 1561:1997	Founding. Grey cast irons
BS EN 50081 (various)	Electromagnetic compatibility. Generic emission standard.
BS EN 50082 (various)	Electromagnetic compatibility. Generic immunity standard.
BS EN 60034 (various)	Rotating electrical machines.

Other references/information sources

The Lifting Operations and Lifting Equipment Regulations (LOLER) 1998

The Lift Regulations 1997

The Provision and Use of Work Equipment Regulations (PUWER) 1998

SAFED	Lift Guidance Note LG 1 (supersedes HSE Guidance Note PM7 Lifts: Thorough examination and testing.)

The Supply of Machinery (Safety) Regulations (with amendments) 1992 (the UK enactment of the EC Lifts Directive 95/16/EC)

PAS 32–1:1999	Specification for examination and testing of new lifts before putting into service. Electric traction lifts.
G5/4	Electricity Association Engineering Recommendation 'Limits for harmonics in the G5/4 Electricity Association Engineering Recommendation 'Limits for harmonics in the UK electricity supply system.

Note: For lifts in housing, LOLER (ACOP) 45, page 11 provides guidance on the Landlord's duties under Health & Safety at Work Act.

Notes on standards

For individual lift components, many manufacturers and their UK agents quote only the standards applicable to the country of origin, and not the relevant British or European standards. However, all UK manufacturers and suppliers and all UK installations must comply with the Life Regulations 1997 and all new installations must bear the CE mark. All installations must also comply with BS EN 81.

ELECTRIC TRACTION LIFTS

	MANUFACTURERS' 'PACKAGED' LIFTS		
YEARS	DESCRIPTION	INSPECTION	MAINTENANCE
15	Complete electric traction lift, geared or gearless, constructed & installed to BS EN 81–1, with sub-components to BS 5655 & CE marking to Lift Regulations 1997.	Periodic thorough inspection and testing by 'competent persons' in accordance with manufacturer's requirements and with the following legislation:	Periodic maintenance by 'competent persons' in accordance with manufacturer's and statutory requirements (see inspection), and in response to feedback from statutory/scheduled inspections.
U	Unclassified i.e. electric traction lift, geared or gearless, not constructed to the above standards.	• Lifting Operations and Lifting Equipment Regulations (LOLER) 1998 • Provision and Use of Work Equipment Regulations (PUWER) 1998 • Lift Regulations 1997 • Supply of Machinery (Safety) Regulations (with amendments) 1992. The LOLER regulations require that inspections of passenger lifts are carried out at a minimum of six monthly intervals. The Safety Assessment Federation (SAFED) Guidelines on the Thorough Examination & Testing of Lifts (LG1) provide detailed recommendations for 1, 5 and 10 yearly inspections. Monthly/three monthly inspections in accordance with manufacturer's and professional adviser's schedules.	

	BESPOKE LIFTS		
YEARS	DESCRIPTION	INSPECTION	MAINTENANCE
	Electric traction lift, geared or gearless, constructed & installed to BS EN 81–1, with sub-components below to BS 5655 & CE marking to Lift Regulations 1997.	As above.	As above.
35+	Car sling, counterweight and guides		
20–25	Lift car		
10–20	Lift car doors (see separate detailed schedule)		
15	Control panel (IC types), call buttons and indicators		
10–20	Drive unit, including motor, gearbox, brake, sheave, drum and bedplate (see separate detailed schedule)		
10–12	Lift ropes (see separate detailed schedule)		
U	Unclassified i.e. electric traction lift, geared or gearless, components not constructed/installed to BS EN 81–1 or BS 5655.		

Adjustment factors

Manufacturing process not quality assured i.e. to ISO 9000 series: –5 years.

Lifts installed in a group operate under group collective control, with the most demanding duties are attributed by rotation: +5 years.

Assumptions – Design & Installation

Six monthly and annual inspections to LG 1 to be carried out by a competent person from an appropriate notified body.

Lift motor room to be adequately ventilated as required by BS 7255. Operating conditions to be maintained within temperature limits as defined by manufacturer.

Access for maintenance to be clear, safe and well lit as required by BS 5655 & BS 7255.

Lift and all sub-components to be CE marked as required by Lift Regulations 1997.

Installation in strict accordance with manufacturer's instructions and relevant BSRIA/CIBSE/HVCA/LEIA other guidance/codes.

All lifts to be assigned a duty rating of light, medium or heavy. Suitable match to be made between expected usage and related duty and the building type in which the lift is installed.

Incorporation of a data logging, fault detection and remote monitoring system (including hours run & number of operations) to BS EN 627 is recommended.

Assumptions – Commissioning

Commissioning in strict accordance with manufacturer's instructions and BSRIA Application Guides/ CIBSE Commissioning Codes/other appropriate guidance

On completion of installation works, commissioning tests & thorough examinations, in accordance with BS 5655 Part 10 should be carried out prior to being handed to the user. The forms detailed in BS 5655 Part 10 are to be completed by a competent person. Note: New procedures will shortly be available.

Maintenance organisations should be a member of the Lift and Escalator Industry Association (LEIA) or similar industry body.

Examination and testing of traction lifts before handover, to PAS 32–1:1999.

Key failure modes

Door, door sub-components and door operators – mechanical or electrical breakdown.

Controllers, contactors, relays, printed circuit, processor boards, solid state devices – electrical breakdown or failure of moving parts (where they are part of the mechanism.)

Computer control – failure.

Ropes: broken strands, migration of internal lubricant, flattening, sheave wear, fatigue, stretching, shock loading or overloading twisting due to poor or unskilled handling, reduction in diameter.

Lift car, display panels, call panels – durability affected by vandalism and interference by passengers.

Motors – vibration, noise or failure – overload, moisture laden or contaminated atmosphere, unstable supply characteristics, wear of bearings.

Car levelling sub-components – poor alignment caused by vibration, damage by passing objects and accidents during maintenance.

Drive-sheave – wear. If wear is uneven or if sheave is of sufficient size and original quality, rope grooves can be re-cut.

Gearbox/gears: wear.

Brakes: wear.

Key durability issues

Frequency of use.

Standard of manufacture and quality of materials, especially if a 'packaged lift' is chosen instead of a 'bespoke lift'.

Quality of installation and commissioning.

Quality of maintenance and inspection regimes and personnel.

It is important to match the tensile strength of the rope to the specification of the traction sheave casting, in order to prevent rapid wear of the casting.

Lang lay ropes typically display up to 30% greater fatigue performance than those with ordinary lay.

Failure of door and door components account for 80% of all lift breakdowns.

Maintenance should be supported and planned using data from monitoring equipment i.e. number of starts, number of calls and running hours.

Vibration monitoring can be used to plan maintenance and reduce noise.

Ropes should not be excessively lubricated as this increases the tendency to slip and increase wear.

LG 1 inspections play a key role in maintaining reliability.

Drive sheave to be as large as possible to reduce stresses in traction ropes i.e. diameter of sheave should be 45 times the diameter of the rope. Note: EN 81 states 40:1 ratio.

Life of lift car very much dependent on tenant use/abuse. Note that refurbishment of lift cars is often more for cosmetic than functional reasons.

Notes

The Lifts Regulations 1997 require conformity by installers of lifts and manufacturers of safety components, and the regulations specify the options available to achieve such conformity.

The regulations have introduced, within the UK, the concept of appointing highly qualified and experienced organisations (notified bodies), to carry out conformity assessment procedures. To date, five such organisations have been appointed. Details are available from Standards and Technical Regulations Directorate, Section 5, at the Department of Trade and Industry. Notified bodies provide the certificate of conformity.

It is the responsibility of the installer to affix the CE marking in the car of the lift and to draw up the declaration of conformity.

The installer must keep copies of the certificate of conformity and the declaration of conformity for a period of 10 years and make them available for inspection.

The newly introduced regulations insist on a much tougher maintenance and testing regime, which in the case of buffers may lead to 'destructive-type testing'. It also requires '24-hour, 2-way voice communication'.

The new testing regulations concern lifts in 'places of work' and only applies to new lifts. However, if an older lift is 'materially changed or its characteristic(s) are changed', then it may be classed as a 'new' lift and subject to the new regulations.

ELECTRIC TRACTION LIFTS *(continued)*

LIFT ROPES

YEARS	DESCRIPTION	INSPECTION	MAINTENANCE
12	Suspension ropes to BS 302:Part 4. Lang lay.	Periodic thorough inspection and testing by 'competent persons' in accordance with manufacturer's requirements and with the following legislation:	Periodic maintenance by 'competent persons' in accordance with manufacturer's and statutory requirements (see inspection), and in response to feedback from statutory/scheduled inspections.
10	Suspension ropes to BS 302:Part 4. Ordinary lay.		
10	Ancilliary ropes (i.e. governor ropes, compensating ropes, door gear ropes) to BS 302:Part 4.		
U	Unclassified, i.e. ropes not in accordance with BS 302:Part 4.		

Legislation:

• Lifting Operations and Lifting Equipment Regulations (LOLER) 1998

• Provision and Use of Work Equipment Regulations (PUWER) 1998

• Lift Regulations 1997

• Supply of Machinery (Safety) Regulations (with amendments) 1992.

The LOLER regulations require that inspections of passenger lifts are carried out at a minimum of six monthly intervals.

The Safety Assessment Federation (SAFED) Guidelines on the Thorough Examination & Testing of Lifts (LG1) provide detailed recommendations for 1, 5 and 10 yearly inspections.

Monthly/three monthly inspections in accordance with manufacturer's and professional adviser's schedules.

Adjustment factors

Gearless drive applications: +2 years (except lang lay ropes).

Single speed lifts with high deceleration rates: –2 years.

Suspension ropes operating in 'V' groove sheaves: –5 years.

Assumptions – Design & Installation

Maintenance to be carried out by a member of the Lift and Escalator Industry Association (LEIA) with ready supply of spares. Data logging, fault detection and remote monitoring system (including hours run & number of operations) to BS EN 627.

Six monthly and annual inspections to LG 1 to be carried out by a competent person from an appropriate notified body.

It is important to match the tensile strength of the rope to the specification of the traction sheave casting, in order to prevent rapid wear of the casting.

Installation in strict accordance with manufacturer's instructions and relevant BSRIA/CIBSE/HVCA/LEIA other guidance/codes.

Particular care should be taken with the installation of lang lay ropes, which are more prone to disturbance of the lay during installation.

Ideally, all the ropes for any one lift should be taken from one production length, in order to eliminate variations in stretch characteristics.

Assumptions – Commissioning

Commissioning in strict accordance with manufacturer's instructions and BSRIA Application Guides/CIBSE Commissioning Codes/other appropriate guidance

On completion of installation works, commissioning tests & thorough examinations, in accordance with BS 5655 Part 10 should be carried out prior to being handed to the user. The forms detailed in BS 5655 Part 10 are to be completed by a competent person. Note: New procedures will shortly be available.

Maintenance to be carried out by suitably qualified and experienced persons. Maintenance organisations should be a member of the Lift and Escalator Industry Association (LEIA) or similar industry body.

Key Failure Modes

Broken strands, migration of internal lubricant, flattening, sheave wear, fatigue, stretching, shock loading or overloading twisting due to poor or unskilled handling, reduction in diameter.

Key Durability Issues

Frequency of use.

Quality of installation and commissioning.

Lang lay ropes typically display up to 30% greater fatigue performance than those with ordinary lay.

'V' groove sheaves can lead to high stresses. Also the 'V' shape can distort the rope shape.

High deceleration rates can increase rope/sheave pressure, particularly on single speed lifts where braking distance is extremely short.

Gearless drive applications exert less pressure on the ropes because the undercut width on the traction sheave is smaller.

Ropes should not be excessively lubricated as this increases the tendency to slip and increase wear.

LG 1 inspections play a key role in maintaining reliability.

Drive sheave to be as large as possible to reduce stresses in traction ropes i.e. diameter of sheave should be 45 times the diameter of the rope. Note: EN 81 states 40;1 ratio.

Notes

The Lifts Regulations 1997 require conformity by installers of lifts and manufacturers of safety components, and the regulations specify the options available to achieve such conformity.

The regulations have introduced, within the UK, the concept of appointing highly qualified and experienced organisations (notified bodies), to carry out conformity assessment procedures. To date, five such organisations have been appointed. Details are available from Standards and Technical Regulations Directorate, Section 5, at the Department of Trade and Industry. Notified bodies provide the certificate of conformity.

It is the responsibility of the installer to affix the CE marking in the car of the lift and to draw up the declaration of conformity.

The installer must keep copies of the certificate of conformity and the declaration of conformity for a period of 10 years and make them available for inspection.

The newly introduced regulations insist on a much tougher maintenance and testing regime, which in the case of buffers may lead to 'destructive-type testing'. It also requires '24-hour, 2-way voice communication'.

The new testing regulations concern lifts in 'places of work' and only applies to new lifts. However, if an older lift is 'materially changed or its characteristic(s) are changed', then it may be classed as a 'new' lift and subject to the new regulations.

ELECTRIC TRACTION LIFTS *(continued)*

YEARS	DESCRIPTION	INSPECTION	MAINTENANCE
	DRIVE UNITS		
20	Variable voltage, variable frequency (VVVF) AC gearless drive to BS 4999, BS 5000, BS EN 60034 and BS EN 81–1.	Periodic thorough inspection and testing by 'competent persons' in accordance with manufacturer's requirements and with the following legislation:	Periodic maintenance by 'competent persons' in accordance with manufacturer's and statutory requirements (see inspection), and in response to feedback from statutory/scheduled inspections.
20	Variable voltage DC gearless drives to BS 4999, BS EN 60034 and BS EN 81–1.		
15	Variable voltage, variable frequency (VVVF) AC geared drive to BS 4999, BS 5000, BS EN 60034 and BS EN 81–1. Gearbox to BS 721.	• Lifting Operations and Lifting Equipment Regulations (LOLER) 1998	
15	Variable voltage DC geared drives to BS 4999, BS EN 60034 and BS EN 81–1. Gearbox to BS 721.	• Provision and Use of Work Equipment Regulations (PUWER) 1998	
15	Variable voltage AC geared drive to BS 4999, BS 5000, BS EN 60034 and BS EN 81–1. Gearbox to BS 721.	• Lift Regulations 1997	
10	Two speed AC motors to BS 4999, BS 5000, BS EN 60034 and BS EN 81–1. Gearbox to BS 721.	• Supply of Machinery (Safety) Regulations (with amendments) 1992.	
10	Single speed AC motors to BS 4999, BS 5000, BS EN 60034 and BS EN 81–1. Gearbox to BS 721.	The LOLER regulations require that inspections of passenger lifts are carried out at a minimum of six monthly intervals.	
U	Unclassified, i.e. drives not to above standards.	The Safety Assessment Federation (SAFED) Guidelines on the Thorough Examination & Testing of Lifts (LG1) provide detailed recommendations for 1, 5 and 10 yearly inspections.	
		Monthly/three monthly inspections in accordance with manufacturer's and professional adviser's schedules.	

Adjustment factors

Manufacturing/installation by a body not quality assured to ISO 9000 series: –5 years (15 and 20 year drives only).

Sealed for life bearings used instead of re-greasable bearings: –5 years (15 and 20 year drives only).

High standard of maintenance supported by condition monitoring: +5 years.

The above factors are not cumulative: the factor that is the largest should be applied.

Assumptions – Design & Installation

Motor protection to BS 4999; IP20 for clean environments and IP40 or better for locations exposed to dust and dirt. Windings to have class F insulation.

Motor to have thermostor temperature protection.

Maximum rated output for AC drives = 80kW approx; DC drives = 200kW approx.

Large frame motors to have forced cooling.

Drives to meet electro-magnetic (EMC) emission and immunity standards of BS EN 50081 and BS EN 50082 respectively.

Storage, unpacking and installation of drives in strict accordance with manufacturers' instructions.

Motor windings to be tested for low insulation values after storage or transit to assess the effects of damp environments.

Traction lift motor rating to be to BS 5655:Part 5 or better.

Lift motor room to be adequately ventilated as required by BS 7255. Operating conditions to be maintained within temperature limits as defined by manufacturer. BS 5655:Part 6 requires machine room ventilation and heating to maintain air temperature at between 15 and 35°C.

Geared drives and motors should ideally be fitted with rolling element re-greasible bearings and not sealed for life or sleeve bearings.

Factory made coupling between motor and gearbox is preferable due to enhanced accuracy.

Hardness of ropes and sheaves to be compatible and wear characteristics of each to be uniform.

Cast iron sheaves to be to BS EN 1561. For ropes, see separate study.

Sheave wear is related to groove pressure; the greater the pressure, the greater the wear. Groove pressure should be no more than 60% of maximum allowable pressure.

Predicted groove pressure may be reduced at design stage by increasing the number of ropes and/or increasing rope diameter.

Compensation ropes required for travel above 30m.

Worm wheel of geared drives to be of phosphor bronze, with good machining properties and high resistance to wear. Sheave and worm wheel to be as large as practicable in order to minimise wear and maximise durability.

Worm gear to BS 721 and BS 5655:Part 6. Note that some parts of BS 721 are not specifically applicable to traction lift installations.

Data logging, fault detection and remote monitoring system (including hours run & number of operations) to BS EN 627.

Six monthly and annual inspections to LG 1 to be carried out by a competent person from an appropriate notified body.

Access for maintenance to be clear, safe and well lit as required by BS 5655 & BS 7255.

Lift and all sub-components to be CE marked as required by Lift Regulations 1997.

Installation in strict accordance with manufacturer's instructions and relevant BSRIA/CIBSE/HVCA/LEIA other guidance/codes.

Assumptions – Commissioning

Commissioning in strict accordance with manufacturer's instructions and BSRIA Application Guides/CIBSE Commissioning Codes/other appropriate guidance

On completion of installation works, commissioning tests & thorough examinations, in accordance with BS 5655 Part 10 should be carried out prior to being handed to the user. The forms detailed in BS 5655 Part 10 are to be completed by a competent person. Note: New procedures will shortly be available.

Examination and testing of traction lifts before handover, to PAS 32–1:1999.

For common information relating to electric motors, see separate motors study.

Proximity of worm wheel teeth to worm gear teeth should be carefully set in accordance with manufacturers' instructions to avoid backlash.

Maintenance to be carried out by suitably qualified and experienced persons. Maintenance organisations should be a member of the Lift and Escalator Industry Association (LEIA) or similar industry body.

Key failure modes

Brush wear on DC motors.

Arcing to commutator of DC motors.

Overloading/overheating of motor.

Brake wear, e.g. due to poor maintenance. Note: brake wear is generally greatest with AC single and two speed drives. With other drive types, braking is provided by the gears and the brakes are used primarily to hold the lift at its destination.

Failure of bearings, e.g. due to wear, inadequate/excessive lubrication, use of unspecified lubricant.

Gearbox failure due to wear, inadequate maintenance, lack of topping up or oil changes.

Rope/sheave wear.

Unpredictable failure of printed circuit boards/electronic sub-components, especially due to prolonged high temperatures (i.e. > 35°C).

Backlash on geared drives: excessive clearance between the worm and the worm wheel, oscillation apparent at levelling.

Vibration of single and two speed AC motor drives during deceleration.

Key durability issues

Standard of manufacture and quality of materials.

Quality of installation and commissioning.

Level of maintenance and competence of maintenance operatives. Maintenance supported by condition monitoring and data logging analysis can enhance durability.

Bearing life is a key determinant of the life of traction drives. Regreasable bearings have a greater durability than sealed-for-life type.

Use of 'V' shaped grooves in main sheaves rather than 'U' shaped can significantly reduce rope wear.

Diameter of main sheave and worm wheel should be as large as practicable. Sheave should be larger than worm wheel. Normal ratio of sheave to worm wheel diameters is 1:2. A good working ratio is 1:1.75. Optimum ratio is 1:1.5.

DC motor brushes should be of optimum hardness to prevent rapid wear, but beware scoring damage to commutator.

Use of synthetic oil in gearboxes can improve the efficiency of worm gear by up to 10%, since synthetic lubricants have a lower coefficient of friction. They also require fewer oil changes.

Geared drives have been assigned a lesser life than gearless due to the additional mechanical load placed on the motor, and because gearless drives tend to have larger motors and suffer from fewer problems. In gearless drives, the sheave is connected directly to the motor, which helps to reduce motor loads.

Notes

All AC drive motors are three phase.

For general information on AC motors refer to separate motors study.

Levelling accuracy of single and two speed AC drives depends upon load and direction of travel; drive mechanism cannot compensate.

In geared drives, sheave should be on the same shaft as worm wheel (also termed crown wheel).

Heat dissipated in lift plant rooms for geared and gearless drives is roughly comparable, but is dependent on duty cycle and type of power control system adopted.

VVVF AC drives have proportionately lower power consumption and greater energy efficiency than other drives types.

DC drives are capable of greater speeds and power than AC drives.

Traction lifts are typically capable of up to 240 starts per hour (ups and downs).

HYDRAULIC LIFTS

Scope

This section provides information on hydraulic lifts used for the transport of people (in the workplace). Data is provided on the major component parts of bespoke lifts (i.e. lifts for which each component is specified separately), and for 'packaged' lifts which are specified as a complete unit. Recent developments in 'gear-room-less' installations are excluded from this study.

It should be noted that the longevity of lift components is a highly complex subject and that for many component parts, replacement will be dictated by statutory inspections. The lives given are therefore indicative only.

The following component sub-types are included within this section:

Standards cited

BS 4999 (various parts)	General requirements for rotating electrical machines
BS 5000 (various parts)	Rotating electrical machines of particular types or for particular applications
BS 5655	Lifts and service lifts
Part 5:1989	Specification for the dimensions of standard lift arrangements
Part 6:1990	Code of practice for selection and installation
Part 7:1983	Specification for manual control devices, indicators and additional fittings
Part 8: 1983	Specification for eyebolts for lift suspension
Part 9:1985	Specification for guide rails
Part 10	Testing and examination of lifts and service lifts
Section 10.2	Hydraulic lifts
Subsection 10.2.1	Commissioning tests for new lifts
Part 12:1989	Recommendations for the installation of new, and the modernisation of, hydraulic lifts in existing buildings
BS 7255: 1989	Code of practice for safe working on lifts
BS EN 81–2: 1998	Safety rules for the construction and installation of lifts – Hydraulic lifts
BS EN 627: 1996	Specification for data logging and monitoring of lifts, escalators and passenger conveyors.

Other references/information sources

The Lifting Operations and Lifting Equipment Regulations (LOLER) 1998

Safe use of lifting equipment: LOLER Approved Code of Practice and Guidance. HSE, 1988.

The Lift Regulations 1997

The Provision and Use of Work Equipment Regulations (PUWER) 1998

The Safety Assessment Federation (SAFED) Lift Guidance Note LG 1 (supersedes older HSE Guidance Note PM7 Lifts: Thorough examination and testing.

The Supply of Machinery (Safety) Regulations (with amendments) 1992 (the UK enactment of the EC Lifts Directive 95/16/EC)

Note: For lifts in housing, LOLER (ACOP) 45, page 11 gives guidance on the Landlords duties under Health & Safety at Work Act

HYDRAULIC LIFTS

MANUFACTURERS' 'PACKAGED' LIFTS

YEARS	DESCRIPTION	INSPECTION	MAINTENANCE
15	Complete hydraulic lift, constructed & installed to BS EN 81 Part 2, with sub-components to BS 5655 & CE marking to Lift Regulations 1997.	Periodic thorough inspection and testing by 'competent persons' in accordance with manufacturer's requirements and with the following legislation:	Periodic maintenance by 'competent persons' in accordance with manufacturer's and statutory requirements (see inspection), and in response to feedback from statutory/scheduled inspections.
U	Unclassified i.e. hydraulic lift, not constructed to the above standards.	• Lifting Operations and Lifting Equipment Regulations (LOLER) 1998 • Provision and Use of Work Equipment Regulations (PUWER) 1998 • Lift Regulations 1997 • Supply of Machinery (Safety) Regulations (with amendments) 1992. The LOLER regulations require that inspections of passenger lifts are carried out at a minimum of six monthly intervals. The Safety Assessment Federation (SAFED) Guidelines on the Thorough Examination & Testing of Lifts (LG1) provide detailed recommendations for 1, 5 and 10 yearly inspections. Monthly/three monthly inspections in accordance with manufacturer's and professional adviser's schedules.	

BESPOKE LIFTS

YEARS	DESCRIPTION	INSPECTION	MAINTENANCE
	Hydraulic lift, constructed & installed to BS EN 81 Part 2, with sub-components below to BS 5655 & CE marking to Lift Regulations 1997.	As for manufacturers' 'packaged' lifts	As for manufacturers' 'packaged' lifts
20–25	Car		
10–20	Car doors, door actuators & door rollers: see separate schedules.		
15	Control panel (IC types), call buttons and indicators.		
10–15	Hydraulic pump, ram, cylinder, oil cooler, return valves etc (see separate detailed schedule).		
U	Unclassified, i.e. hydraulic lifts not to BS EN 81 or BS 5655.		

Adjustment factors

Manufacturing process not quality assured i.e. to ISO 9000 series: –5 years.

Lifts installed in a group operate under group collective control and the most demanding duties are attributed by rotation: +5 years

Assumptions – Design & Installation

Six monthly and annual inspections to be carried out by a notified body. Data logging, fault detection and remote monitoring system (including hours run & number of operations) to BS EN 627.

Six monthly and annual inspections to LG 1 to be carried out by a competent person from an appropriate notified body.

Lift motor room to be adequately ventilated as required by BS 7255 and temperature regulated for oil reservoir. Operating conditions to be maintained within temperature limits as defined by manufacturer.

Access for maintenance to be clear, safe and well lit as required by BS 5655 & BS 7255.

Lift and all sub-components to be CE marked as required by Lift Regulations 1997.

Temperature of hydraulic oil in reservoir to be contained within limits as defined by manufacturer i.e. if temperature too low or high, the corresponding viscosity of the oil affects lift performance.

Installation in strict accordance with manufacturer's instructions and relevant BSRIA/CIBSE/HVCA/LEIA other guidance/codes.

All lifts to be assigned a duty rating of light, medium or heavy. Suitable match to be made between expected usage and related duty and the building type in which the lift is installed.

If the installation is to be used intensively, it may be necessary to provide an oil cooler and air conditioning in the lift machine room.

Data logging, fault detection and remote monitoring system (including hours run & number of operations) to BS EN 627.

Assumptions – Commissioning

Commissioning in strict accordance with manufacturer's instructions and BSRIA Application Guides/CIBSE Commissioning Codes/other appropriate guidance

On completion of installation works, commissioning tests & thorough examinations, in accordance with BS 5655 Part 10 should be carried out prior to being handed to the user. The forms detailed in BS 5655 Part 10 are to be completed by a competent person. Note: New procedures will shortly be available.

Maintenance to be carried out by suitably qualified and experienced persons. Maintenance organisations should be members of the Lift and Escalator Industry Association (LEIA) or similar industry body.

Key failure modes

Door, door sub-components and door operators – mechanical or electrical breakdown.

Controllers, contactors, relays, printed circuit, processor boards, solid state devices – electrical breakdown or failure of moving parts (where they are part of the mechanism)

Computer control – failure.

Lift car, display panels, call panels – durability affected by vandalism and interference by passengers.

Car levelling sub-components; poor alignment caused by vibration, damage by passing objects and accidents during maintenance.

Ram: if a sub-standard ram is specified, it may deflect too much and cause excessive wear to itself or to the sleeve.

Key durability issues

Standard of manufacture and quality of materials, especially if a packaged lift is chosen instead of a bespoke lift.

Quality of installation and commissioning.

Quality of maintenance and inspection regimes and personnel.

Failure of door and door components account for 80% of all lift breakdowns. Maintenance to be supported and planned using data from monitoring equipment i.e. number of starts, number of calls and running hours.

Vibration monitoring can be used to plan maintenance and reduce noise.

LG 1 inspections play a key role in maintaining reliability.

Notes

The Lifts Regulations 1997 require conformity by installers of lifts and manufacturers of safety components, and the regulations specify the options available to achieve such conformity.

The regulations have introduced, within the UK, the concept of appointing highly qualified and experienced organisations, notified bodies, to carry out conformity assessment procedures. Details are available from Standards and Technical Regulations Directorate, Section 5, at the Department of Trade and Industry. Notified bodies provides the certificate of conformity.

It is the responsibility of the installer to affix the CE marking in the car of the lift and to draw up the declaration of conformity.

The installer must keep copies of the certificate of conformity and the declaration of conformity for a period of 10 years and make them available for inspection.

On completion of installation works, commissioning tests & thorough examinations, in accordance with BS 5655 Part 10 should be carried out prior to being handed to the user. The forms detailed in BS 5655 Part 10 are to be completed by a competent person. Note: New procedures will be shortly available.

The new testing regulations concern lifts in 'places of work' and only applies to new lifts. However, if an older lift is 'materially changed or its characteristic(s) are changed', then it may be classed as a 'new' lift and subject to the new regulations.

HYDRAULIC LIFTS

	DRIVES		
YEARS	DESCRIPTION	INSPECTION	MAINTENANCE
35+	Hydraulic pipework – rigid steel to BS 3602:Part 1 and BS EN 81–2.	Periodic thorough inspection and testing by 'competent persons' in accordance with manufacturer's requirements and with the following legislation:	Periodic maintenance by 'competent persons' in accordance with manufacturer's and statutory requirements (see inspection), and in response to feedback from statutory/scheduled inspections.
15	Hydraulic pump – submersible to BS EN 81–2.		
15	Hydraulic pump motor – submersible to BS EN 81–2, BS 4999, BS 5000, IEC 34.		
15	Hydraulic ram/jack and cylinder to BS EN 81–2.	• Lifting Operations and Lifting Equipment Regulations (LOLER) 1998	
15	Hydraulic valves and valve block to ISO CD 1144, ISO 10770, BS EN 81–2.		
7	Hydraulic pipework – flexible, wire braided rubber to BS EN 853, BS EN 81–2. (Low pressure pipes may be of other materials and to other standards).	• Provision and Use of Work Equipment Regulations (PUWER) 1998	
5–7	Valve seals to BS EN 81–2.	• Lift Regulations 1997	
2–4	Ram/jack seals to BS EN 81–2.	• Supply of Machinery (Safety) Regulations (with amendments) 1992.	
U	Unclassified, i.e. hydraulic lift components not to above standards.	The LOLER regulations require that inspections of passenger lifts are carried out at a minimum of six monthly intervals.	
		The Safety Assessment Federation (SAFED) Guidelines on the Thorough Examination & Testing of Lifts (LG1) provide detailed recommendations for 1, 5 and 10 yearly inspections.	
		Monthly/three monthly inspections in accordance with manufacturer's and professional adviser's schedules.	

Adjustment factors

Manufacturing/installation by a body not quality assured to ISO 9000 series: –5 years (pump, motor, ram/jack, valves only).

System operating at temperature in excess of 35°C: –5 years (pump, motor only).

High standard of maintenance supported by condition monitoring: +5 years (pump, motor, ram/jack, valves only).

Note: the above factors are not cumulative: the factor that is the largest should be applied.

Assumptions – Design & Installation

Not all of the major sub-components of hydraulic lift installations are fully covered at present by British or European standards. Some manufacturers cite only the standards applicable to the country of manufacture, whilst others cite compliance with overall lift standards BS EN 81and/or BS 5655.

Control valve rating and performance, and motor rating to BS 5655-6.

Pump motor to be rated for predicted number of motor starts and for predicted duty with at least 10% spare capacity.

Valve systems to BS 5655-6 or better to provide highest economical standard of ride comfort and leveling.

Manufacturer/supplier to be informed of client's preference for mineral or synthetic fluid.
(see Assumptions – Commissioning).

Flexible hydraulic pipes to be date stamped with manufacturer's test date to BS EN 81-2. Storage life before use to be noted at time of installation, and recommended storage times not to be exceeded.

Selection of the appropriate seal material depends upon the speed of operation, diameter of ram, operating temperature and hydraulic fluid type. The most common seal materials are polyurethane, PTFE, and NBR (acrylonitrile butadiene rubber).

For indirect drive lifts the hardness of ropes and sheaves is to be compatible and the wear of each should be uniform.

For heavy duty indirect drives see 'Assumptions – Design & Installation' for traction lifts.

Data logging, fault detection and remote monitoring system (including hours run & number of operations) to BS EN 627.

Six monthly and annual inspections to LG 1 to be carried out by a competent person from an appropriate notified body.

Lift motor room to be adequately ventilated as required by BS 7255. Operating conditions to be maintained within temperature limits as defined by manufacturer. This will often necessitate provision of oil cooling (air blown or recirculating water types).

Access for maintenance to be clear, safe and well lit as required by BS 5655 & BS 7255.

Lift and all sub-components to be CE marked as required by Lift Regulations 1997.

Installation in strict accordance with manufacturer's instructions and relevant BSRIA/CIBSE/HVCA/LEIA other guidance/codes.

Assumptions – Commissioning

Commissioning in strict accordance with manufacturer's instructions and BSRIA Application Guides/CIBSE Commissioning Codes/other appropriate guidance

On completion of installation works, commissioning tests & thorough examinations, in accordance with BS 5655 Part 10 should be carried out prior to being handed to the user. The forms detailed in BS 5655 Part 10 are to be completed by a competent person. Note: New procedures will shortly be available.

Maintenance to be carried out by suitably qualified and experienced persons. Maintenance organisations should be members of the Lift and Escalator Industry Association (LEIA) or similar industry body.

Factory pre-delivery tests are to be executed with the same type of hydraulic fluid as that with which the system is to be filled. Mineral oils should not be used to test systems that will be operated with synthetic oils; synthetic oils will mix with mineral oils even where residues are small.

Submersible pumps are to be covered by the oil at all times, viz. when ram fully extended. The oil tank should hold a minimum of 110% of the system fluid and must be capable of accommodating all of the fluid when the system is drained down.

Key failure modes

Motor failure, e.g. due to overheating. Under-sizing rating of motors is a common cause of premature failure.

Pump failure, e.g. due to overheating.

Fouling of hydraulic fluid through ingress of dirt at sealed joints.

Poor performance due to high temperature of hydraulic fluid.

Incorrect specification of hydraulic fluid.

Valve failure.

Value actuator failure.

Ropes on indirect installations not checked and replaced according to maintenance schedule.

Corrosion of ram/jack. Note that where the base of the cylinder is located within a bore hole below the lift pit, corrosion can occur undetected.

Leaking of hydraulic fluid.

Key durability issues

Standards of manufacture and quality of materials not to recognized British, European or international standard.

Package lift drive accepted in place of bespoke drive.

Quality of installation and commissioning.

Quality maintenance and inspection regime to be executed by competent persons with condition monitoring.

Provision of temperature control to plant room, including oil cooling where necessary. Oil temperature control can greatly extend oil and seal life and enhance the durability of pumps and valves.

Pump motor rated with 10% spare capacity.

Maintenance supported by condition monitoring and data logging analysis. Data logging to BS EN 627 to predict planned replacement of less durable sub-components.

Use of synthetic fluid in place of mineral oil can reduce wear. Oil filtration necessary to limit contamination and maintain purity.

Six monthly sampling and test of hydraulic fluid is recommended as part of condition monitoring regime.

Notes

Hydraulic fluid change – 1 to 2 yearly for mineral oil and 2 to 4 yearly for synthetic depending on site conditions and test sample assessment.

ISO VG 46 viscosity fluid recommended for normal use and ISO VG 63 for heavy use.

Vegetable (organic) oils to be avoided. Shelf and working life limited and oil decomposes. Vegetable oil must not be used to ease seal mounting.

Certain seal materials have limited shelf life (only 6 months). Seals should be ordered in advance of use but not stocked.

Average starts per hour 30; maximum starts per hour 45 – ups only.

LIFT CAR DOORS

Scope

This component study considers common types of lift car doors for passenger lifts in non-domestic building types.

The following are not included in this study: Lift doors made of glass, lift doors to be installed in external locations or exposed to weather.

Standards cited

BS 5655	Lifts and service lifts (various parts).
BS 7255: 1989	Code of practice for safe working on lifts
BS EN 81	Safety rules for the construction and installation of lifts
Part 1:1998	Electric lifts
Part 2:1998	Hydraulic lifts
BS EN 627:1996	Specification for data logging and monitoring of lifts, escalators and passenger conveyors.

Other references/information sources

The Lifting Operations and Lifting Equipment Regulations (LOLER) 1998

The Lift Regulations 1997

The Provision and Use of Work Equipment Regulations (PUWER) 1998

The Safety Assessment Federation (SAFED) Lift Guidance Note LG 1 (supersedes older HSE Guidance Note PM7 Lifts: Thorough examination and testing).

The Supply of Machinery (Safety) Regulations (with amendments) 1992 (the UK enactment of the EC Lifts Directive 95/16/EC)

LIFT CAR DOORS

LIFT CAR DOORS

YEARS	DESCRIPTION	INSPECTION	MAINTENANCE
20	Stainless or galvanized steel faced, single panel, side opening door to BS 5655, BS EN 81 and Lift Regulations 1997 with ac or dc motor and toothed drive belt. Optical sensor-controlled obstacle detection/door opening.	Periodic thorough inspection and testing by 'competent persons' in accordance with manufacturer's requirements and with the following legislation:	Periodic maintenance by 'competent persons' in accordance with manufacturer's and statutory requirements (see inspection), and in response to feedback from statutory/scheduled inspections.
15	Stainless or galvanized steel faced, two panel, centre opening door to BS 5655, BS EN 81 and Lift Regulations 1997 with ac or dc motor and toothed drive belt. Optical sensor controlled obstacle detection	• Lifting Operations and Lifting Equipment Regulations (LOLER) 1998	Maintenance to include (typically): *Weekly:* Vacuum, (not sweep) door sills.
10	Stainless or galvanized steel faced, four panel, centre opening door to BS 5655, BS EN 81 and Lift Regulations 1997 with ac or dc motor and toothed drive belt. Optical sensor controlled obstacle detection	• Provision and Use of Work Equipment Regulations (PUWER) 1998 • Lift Regulations 1997	*Three monthly:* Check and clean door rollers. Check and clean door hanger tracks.
U	Unclassified, i.e. doors not to BS 5655, BS EN 81 or Lift Regulations 1997.	• Supply of Machinery (Safety) Regulations (with amendments) 1992. The LOLER regulations require that inspections of passenger lifts are carried out at a minimum of six monthly intervals. The Safety Assessment Federation (SAFED) Guidelines on the Thorough Examination & Testing of Lifts (LG1) provide detailed recommendations for 1, 5 and 10 yearly inspections. Monthly/three monthly inspections in accordance with manufacturer's and professional adviser's schedules.	*Annually:* Clean, check and tighten electrical terminations and connections. *Five yearly:* Replace door shoes & safety edges. *Eight to ten yearly:* Replace drive belts and motor. *Ten to fifteen yearly:* Replace door rollers and top tracks. Re-skim doors where necessary.

Adjustment factors

Obstacle detection and door opening triggered by optical sensors rather than traditional electro-mechanical leading edges: +5 years.

Assumptions – Design and Installation

Installation in strict accordance with manufacturers' and professional advisors' requirements.

Inspections to LG 1 to be carried out by a competent person from an appropriate notified body.

Lift doors to be maintained to manufacturers' and professional advisors' requirements.

Sub-components to be CE marked as required by Lift Regulations 1997.

Panelled doors to comprise mild steel frame with stainless steel skin of 1.5 mm thickness and to be capable of withstanding impact loads perpendicular to the surface as required by BS 5655 Parts 1 & 2.

Landing door interlocks to be subjected to annual inspection and certification as detailed within SAFED Lift Guidelines LG1. Further inspection and certification to be part of five and ten yearly procedures as detailed in LG1.

Where time elapsed maintenance is shown in the table above, manufacturers to provide predictive maintenance/replacement information related to hours in service, number of trips or number of operations.

Lift and all sub-components to be CE marked as required by Lift Regulations 1997.

For optimal performance, door hanger rollers should be not less than 80 mm diameter with neoprene tyres to reduce wear.

Doors may be single or double skinned, with or without a noise absorbent lining, depending on environment and duty, viz. heavy industrial production unit or high quality office accommodation. All versions are likely to be equally durable in their selected environment. Very heavy doors are slow to operate and can place additional loads on the door operating system.

Door acceleration and deceleration during opening and closing operations to run smoothly and not by step changes in speed.

Incorporation of data logging, fault detection and remote monitoring system (including hours run & number of operations) to BS EN 627, is recommended.

Assumptions – Commissioning

Commissioning in strict accordance with manufacturer's instructions and BSRIA Application Guides/CIBSE Commissioning Codes/other appropriate guidance.

On completion of installation works, commissioning tests & thorough examinations, in accordance with BS 5655 Part 10 should be carried out prior to being handed to the user. The forms detailed in BS 5655 Part 10 are to be completed by a competent person. Note: New procedures will shortly be available.

Maintenance to be carried out by suitably qualified and experienced persons. Maintenance organisations should be a member of the Lift and Escalator Industry Association (LEIA) or similar industry body.

Key failure modes

Door, door sub-components and door operators – mechanical or electrical breakdown.

Impact damage to doors.

Impact damage to mechanical leading edges – if installed.

Damage to door shoes, rollers and door sill debris

Key durability issues

Standard of manufacture and quality of materials, especially if a 'packaged lift' is chosen instead of a 'bespoke lift'.

Mechanical safety edges to doors are frequently subjected to impact damage either through passenger misuse or due to doors striking hard objects in their path. Electronic safety edges detect obstacles using a complex of infrared beams and reverse the direction of the doors before any contact is made with the obstruction. There is therefore no contact with the safety edge and no impact shock to the door operator.

Heavier gauge door panels are better able to withstand impact damage.

Use of doors with an impressed pattern can reduce the need for cosmetic cleaning and can reduce apparent surface deterioration by small scratches and knocks.

Manually operated swing doors are not considered as reliable as automatic doors because of the limited reliability of the door locking mechanism.

Failure of doors and door components account for 80% of all lift breakdowns. Maintenance should be supported and planned using data from monitoring equipment i.e. number of starts, number of calls and running hours.

Notes

The heavier the gauge of the materials used in door manufacture, the more durable the door is likely to be, but the slower the opening and closing speeds required, in order to comply with the requirements of BS 5655 Parts 1 & 2.

Clear entrance dimensions and internal car dimensions are contained in BS 5655 Part 5. Equivalent information for service lifts is contained in BS 5655 Part 3.

Lift car door operating mechanisms require more frequent maintenance attention than landing door operating mechanisms. The most frequently used landing doors (e.g. ground floor entrance hall) should be identified for additional maintenance.

The Lifts Regulations 1997 require conformity by installers of lifts and manufacturers of safety components, and the regulations specify the options available to achieve such conformity.

The regulations have introduced, within the UK, the concept of appointing highly qualified and experienced organisations (notified bodies), to carry out conformity assessment procedures. To date, five such organisations have been appointed. Details are available from Standards and Technical Regulations Directorate, Section 5, at the Department of Trade and Industry. Notified bodies provide the certificate of conformity.

It is the responsibility of the installer to affix the CE marking in the car of the lift and to draw up the declaration of conformity.

The installer must keep copies of the certificate of conformity and the declaration of conformity for a period of 10 years and make them available for inspection.

The newly introduced regulations insist on a much tougher maintenance and testing regime, which in the case of buffers may lead to 'destructive-type testing'. It also requires '24-hour, 2-way voice communication'.

The new testing regulations concern lifts in 'places of work' and only applies to new lifts. However, if an older lift is 'materially changed or its characteristic(s) are changed', then it may be classed as a 'new' lift and subject to the new regulations.

Alphabetical index
Appendices

Alphabetical index of components

Appendix A: Abbreviations used in the Manual

ACOP   Approved Code of Practice

AHU   Air handling unit

ASHRAE   American Society of Heating, Refrigeration and Air Conditioning

BASEC   British Appraisal Service for Electric Cables

BBA   British Board of Agr(ment

BEMS   Building energy management system

BEWA   British Effluent and Water Association

BRE   Building Research Establishment

BSI   British Standards Institute

BSRIA   Building Services Research and Innovation Association

CIBSE   Chartered Institute of Building Services Engineers

CORGI   Council of Registered Gas Installers

DEO   Defence Estates Organisation

FMA   Fan Manufacturers' Association

HAPM   Housing Association Property Mutual (latent defects insurance scheme)

HSE   Health and Safety Executive

HTHW   High temperature hot water

HVAC   Heating, ventilation and air conditioning

HVCA   Heating and Ventilation Contractors' Association

IEE   Institute of Electrical Engineers

IOP   Institute of Plumbing

IOR   Institute of Refrigeration

ISO   International Standards Organisation

LEIA   Lift and Escalator Industry Association

LOLER   Lifting Operations and Lifting Equipment Regulations (1998)

LPC   Loss Prevention Council

LPS   Loss Prevention Standard

LTHW   Low temperature hot water

MoD   Ministry of Defence

MTHW   Medium temperature hot water

PSA   Property Services Agency

SAFED   Safety Assessment Federation

WRc   Water Research Council

Appendix B: Definitions for exposure conditions and adjustment factors

The following definitions are intended to aid users of the Manual to determine whether any adjustment factors apply to the particular buildings or locations with which they are concerned.

"Normal environment"

Unless stated otherwise, life assessments are provided on the basis of a normal environment, which is assumed to be inland, with normal urban atmospheric pollution only.

"Polluted/industrial environment"

An environment with airborne sulphur dioxide, acid or alkali pollution, normally from an industrial source. The UK Ministry of Agriculture, Fisheries and Food publishes a map every five years showing the average atmospheric corrosivity rate for 10 km grid squares of the UK. This should be taken as the basis for the assessment, with no adjustment made for microclimatic differences.

"Marine environment"

Coastal areas subject to salt spray and/or sea water splashes. These may extend up to 3 km from the coast or tidal estuary depending on prevailing wind and topography.

"Adverse (but not severely corrosive) environment"

An internal environment exposed to dampness, humidity or mild chemical pollution; an external location with direct exposure to the elements or to airborne pollution.

"Damp/humid environments"

Internal environments subject to frequent and/or severe wetting or condensation, eg kitchens, bath/shower/changing rooms, laundries, swimming pool enclosures.

"With 3rd party assurance"

A product with a certificate indicating that ongoing testing and assessment of the product's suitability and/or adherence to claimed standards is carried out by an independent third party, e.g. British Board of Agrément certification, BSI Kitemarking. Note that BS EN ISO 9000 certification does not match this definition.

"Normal operating hours"

Unless stated otherwise, plant operating hours are assumed to be based on 12 hours per day, five days per week, throughout the year.